T0135761

From the Institute of Signal Processing
of the University of Lübeck
Director: Prof. Dr.-Ing. Alfred Mertins

Invariant Features and Enhanced Speaker Normalization for Automatic Speech Recognition

Dissertation
for Fulfillment of
Requirements
for the Doctoral Degree
of the University of Lübeck

from the Department of Computer Science/Engineering

Submitted by

Florian Müller
from Lübeck

Lübeck, July 2012

Bibliografische Information der Deutschen Nationalbibliothek

Die Deutsche Nationalbibliothek verzeichnet diese Publikation in der Deutschen Nationalbibliografie; detaillierte bibliografische Daten sind im Internet über http://dnb.d-nb.de abrufbar.

ISBN 978-3-8325-3319-9

Logos Verlag Berlin GmbH
Comeniushof, Gubener Str. 47,
10243 Berlin
Tel.: +49 (0)30 42 85 10 90
Fax: +49 (0)30 42 85 10 92
INTERNET: http://www.logos-verlag.de

Reviewer:	Prof. Dr.-Ing. Alfred Mertins
	Prof. Dr.-Ing. Erhardt Barth
Dean:	Prof. Dr. rer. nat. Thorsten M. Buzug
Date of oral examination:	6. December 2012

Acknowledgments

Das Erstellen dieser Arbeit wäre ohne eine Reihe von Menschen so nicht möglich gewesen. An erster Stelle möchte ich meinem Doktorvater Herrn Prof. Dr.-Ing. Alfred Mertins danken. Die Diskussionen und Anregungen, sowie seine große Unterstützung in den vielen Bereichen des wissenschaftlichen Umfelds und die Freiheiten, die einem vertrauensvoll zuteil gelassen werden, habe ich sehr zu schätzen gelernt.

Seitens meiner Arbeitskollegen geht mein Dank zuerst an Alexandru Condurache, der zu einem besonderen Anteil zu meinem eingeschlagenen Weg beigetragen hat. Mein Dank gilt an dieser Stelle aber natürlich auch meinen weiteren Kollegen, mit denen ich die Zeit nicht nur im Institut verbringen durfte. Namentlich seien hier Ulrich Hofmann, Radoslaw Mazur, Ole Jungmann, Matthias Pohl, Simon Vogt, Olaf Christ und Dierck Matern in ihrer ganz einzigartigen Zusammensetzung genannt. Radek danke ich auch (aber nicht nur!) für das Aufrechterhalten unserer Rechenkapazitäten und für seine Mühen kurz vor Abgabe dieser Arbeit. Ole und Christian Kier danke ich auch für eine angenehme Zeit im selben Büro. Auch danke ich Cristina Darolti und Heiko Hansen, die schneller als ich davon überzeugt waren, dass diese Arbeit zustande kommt. Ich bin froh über die Erfahrungen, die ich mit den Studenten Eugene, Peter, Mark, Douglas und Jan machen durfte.

Die Deutsche Forschungsgemeinschaft hat die Projekte, an denen ich gearbeitet habe, finanziert, wofür ich ebenfalls sehr dankbar bin.

Natürlich danke ich Katharina, die mein Leben auf so viele Weisen bereichert.

Zuletzt gebührt der Dank noch meiner Familie, die all dies erst möglich gemacht hat.

Contents

1 Introduction

In our daily life we easily make conversations with many different people in many different environments. The ability to hear and understand human speech is a central element for this interactivity. It has evolved for millions of years and the necessary actions for its execution are generally performed unconsciously. This work is about *automatic speech recognition* (ASR). ASR is the task of transcribing a speech signal into its corresponding textual representation by a machine. It is part of the broad field of speech and language processing and a basic component of many real-world applications that have emerged. Examples are transcriptions of news bulletins, information retrieval, translation, and voice search (Feng et al., 2012).

Lippmann (1997) compared the performance of human and machine recognition on six different speech datasets, so-called *corpora*, with vocabularies ranging from 10 to more than 65,000 words and speaking styles being isolated words, read speech, and spontaneous speech. He showed that human recognition performance far exceeds that of machines and is often magnitudes better than recognizers at that time in both quiet and degraded environments. Benzeghiba et al. (2007) argue that even though the experiments of Lippmann have been conducted more than one decade earlier, the gap between machine and human performance is still substantial. With respect to current state-of-the-art ASR systems, Feng et al. (2012) point to the fact that with ASR technology finding its way into everyday applications many interesting challenges of this field still remain or are even created. Speaker-independent feature extraction, noise-robustness, and auditory representations are currently trending fields of research.

In general, human listeners are able to understand speech reliably under a wide range of environmental conditions and various speaker-specific variabilities. Different physical sizes of the speakers represent one variability among these. The speech signal typically reaches the auditory system of a listener in form of a sound wave and passes through several processing stages so that the contained message can be recognized. This means, that the representation of the speech signal within the auditory system is transformed at some point. The result of this transform is independent of the variabilities that have a distorting influence on the speech signal. The "front-end" of a speech recognition system also transforms the input speech signal. It parametrizes the signal into a representation that is well suited for the succeeding recognition by a machine. This stage is also referred to as the feature extraction stage. Current standard feature extraction methods for ASR handle the effects of differently sized speakers commonly not on the feature extraction stage, but on the acoustic modeling and decoding stage. The methods presented in this thesis focus on the feature extraction stage of an ASR system. The involved methods make use of so-called "invariance transforms" in order to obtain robustness against variabilities that are introduced by different speakers. Furthermore, methods for increasing the robustness of a recognizer under noisy conditions are presented. In another chapter two different approaches for estimating the spectral effects due to different vocal tract lengths are presented and used for enhancing a subsequent recognition step.

Research in the field of ASR is interdisciplinary and involves several fields of expertise, which include signal processing, optimization, machine learning, and statistical techniques, as well as linguistics, psychoacoustics, and phonology. The next two sections give an introduction about human speech production and perception, as well as an overview of variabilities that can be observed in the context of speech recognition and that have to be considered for the design of an ASR system. At the end of this chapter an overview of the main contributions of this thesis is given.

1.1. Speech Production and Perception Process

Spoken language is used to communicate information from a speaker to a listener (Huang et al., 2001b). The processes of generating and processing speech can be described as a sequence of activities as shown in Figure 1.1. Though not explicitly shown, the speaker also uses its own speech to get feedback and, therefore, to control the whole speech generating process.

Schematic Overview of Speech Production and Perception The production process is initiated by the speaker's intention to transmit information via speech. A message has to be mentally formulated, which involves semantical and pragmatical information. After its formulation the message has to be converted into a sequence of phonemes. A phoneme can be thought of as an ideal sound unit with a complete set of articulatory gestures (Deller et al., 1993). Certain groups of phonemes represent words and, thus, whole sentences are constructed by stringing together phonemes. The speaker has to decide for the language he wants to use and, therefore, must have lexical and syntactical (which also means grammatical) knowledge about the intended language. Furthermore, the speaker has to choose a certain prosody. Generally, the speaker has to choose a language code for the message in mind. Then the speaker has to perform a series of neuromuscular actions to produce vocal cord vibrations, air pressure, vocal tract shapes, lips, jaw, tongue, and velum movements with an adequate timing. The resulting speech sound wave that is emitted by the speaker is transmitted by a channel.

During the transmission different kinds of distortions occur and interfere with the speech signal. When the speech signal reaches a listener it is passed to the basilar membrane within the inner ear. A nonlinear spectral analysis is performed whose output is converted into activity signals on the auditory nerve. These signals are translated into a language code at higher stages within the listener's brain. Using semantical and pragmatical knowledge, a message comprehension can be achieved by the listener.

Articulators and Model As mentioned above speech production is a dynamic process that involves several anatomical structures. An overview about these structures is given in Figure 1.2. The lungs generate air pressure. This leads to air flowing through the trachea and the pharyngeal, oral, and nasal cavities. The pharyngeal and oral cavities together are commonly referred to as the *vocal tract*. Its length, measured as the distance from the glottis to the lips, will be referred to as *vocal tract length* (VTL) in the following. The *larynx* contains the *vocal folds*. The vocal tract contains the velum, tongue, and the teeth, and is terminated at the lips. Velum, tongue, teeth, and lips are also called *articulators*. From an engineering perspective the vocal tract makes up the main acoustic filter for the production of speech sounds. To produce a voiced speech signal, air is forced through the *glottis*, which is an opening between the vocal folds, such that they oscillate, thereby producing periodic pulses. This leads to harmonic structures in the spectrum of

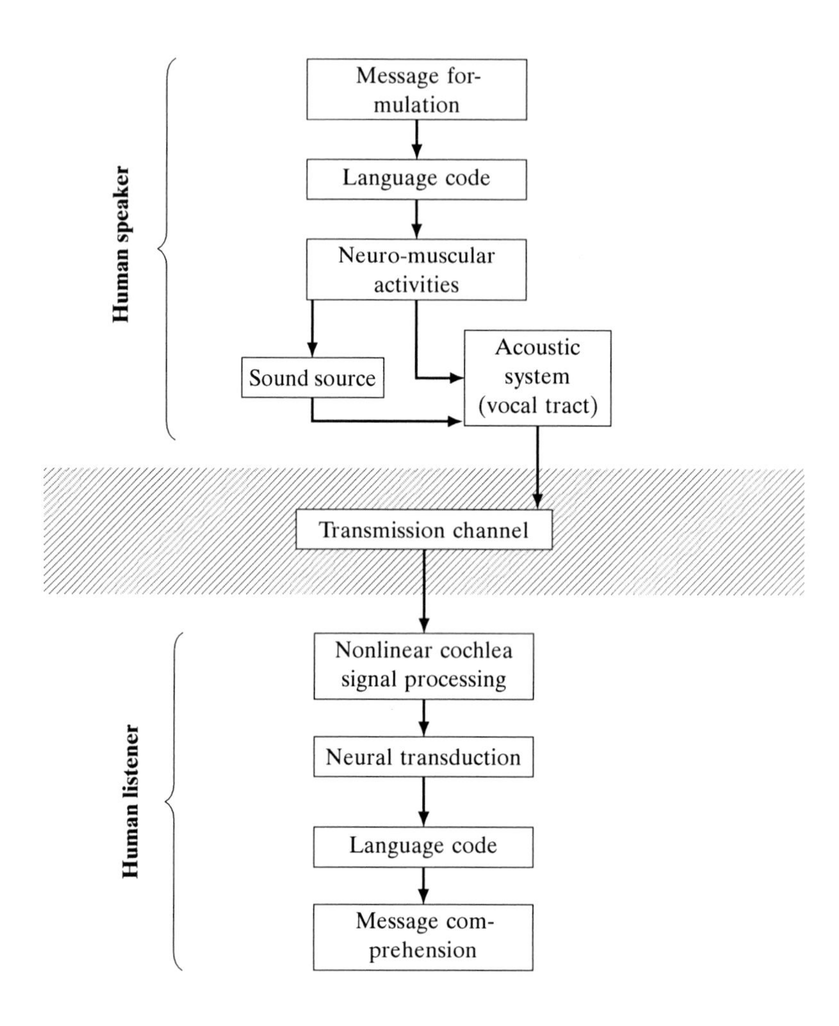

Figure 1.1.: Schematic diagram of the human speech production and perception process. Adapted from Rabiner and Juang (1993); Huang et al. (2001b)

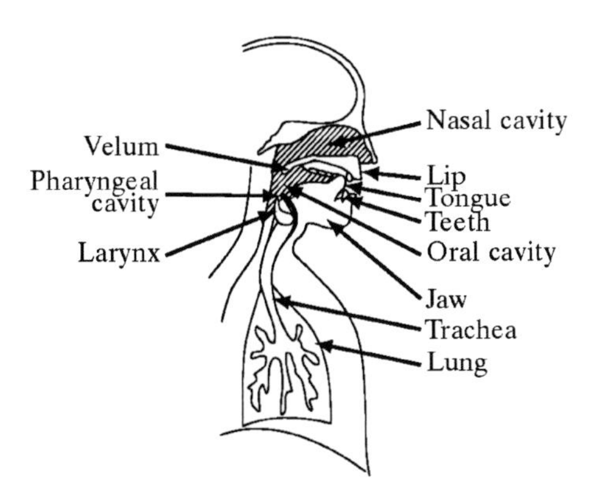

Figure 1.2.: An overview of the articulators for the human speech production (after Deller et al., 1993).

voiced sounds. The rate f_0 of successive openings of the vocal folds is called the *fundamental frequency* or *pitch*. As pointed out in Deller et al. (1993, p. 114), in the field of psychoacoustics the term "pitch" also often refers to the *perceived* fundamental frequency. The characteristics of the system are controlled by the articulators. Unvoiced sounds are generated without the oscillation of the vocal folds and, thus, do not have harmonic structures in their spectrum, and can be characterized by band-limited noise.

A basic discrete-time model for speech production is shown in Figure 1.3. Besides a vocal tract model $H(z)$ a radiation model $R(z)$ is shown here that models the acoustic properties of the transition from the vocal tract to the environment. The whole system is excited by an excitation signal $u(m)$. In case of unvoiced speech intervals the signal is given in form of random noise. A sequence of glottal pulses is given in case of voiced speech intervals. The pulses are modeled with a glottal pulse model $G(z)$ driven by an impulse train with a certain pitch P_0. The gain determines the amplitude of the produced speech signal.

1.2. Variabilities in Human Speech

Generally, today's state-of-the-art ASR systems do not reach the human recognition performance. It is possible to build an ASR system for a specific task in a specific context for a specific speaker that yields a high accuracy. However, due to the

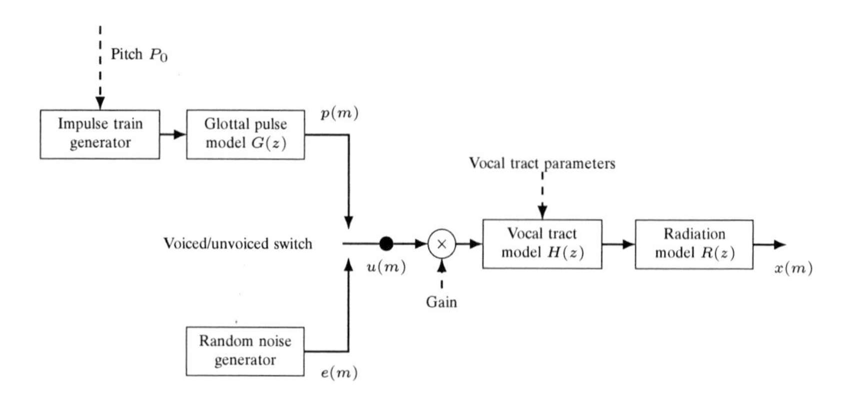

Figure 1.3.: Simplified model for discrete-time human speech production. Adapted from Deller et al. (1993).

amount of variabilities that is inherently present in speech it is still a challenge to build an ASR system that is capable of understanding any language in any environment of every speaker. A comprehensive overview of variabilities that an ASR system has to handle is given in the work of Benzeghiba et al. (2007) and also in the thesis of Stemmer (2005). A broad classification of major variability factors is also given by Huang et al. (2001b, pp. 414-417). A summary about the different types of variabilities is given in the following. No unique classification of variabilities in the field of ASR exists. Here, the different kinds of variabilities are grouped into four categories.

Environmental Variabilities Most applications involve an environment that contains several different types of sound sources. There may be, for example, people talking in the background, machines emitting noise, or bells ringing. Within a room there might be a high amount of reverberation. The distorting noise sources might be standing still or moving and thereby affect the spectral characteristics of the distorting signals. The noise or speech-of-interest characteristics might also change in time when the user is moving. In case of telephone speech, distortions might also originate from different sources, such as from the transmission channel characteristics or from the analog-digital conversion. The microphone(s) of an ASR system will capture all these sources in addition to the user's speech signal.

Variabilities in the Style of Speech Depending on the task or the situation, human speakers articulate in different ways that all have their own characteristics; recognition systems have to handle these characteristics if they are intended to be used under these conditions. If an ASR system is supposed to only recognize isolated words, for example command words, the resulting pauses between the individual commands would provide clear boundaries and silence segments. However, in order to process read speech, such as in broadcast news or dictation tasks, a continuous speech recognizer is needed. The advantage of an ASR system that is capable to understand continuous speech is that it provides a more natural way of interaction with human users. Alleva et al. (1998) reports that users tend to fall back to isolated speech when they feel that the ASR system is not understanding them. However, isolated speech is generally not as well handled by a continuous speech recognition system as on an dedicated isolated speech recognizer. Some reasons for this are different styles of articulation and speaking rates. Read speech and similar styles are usually clearly and consistently articulated.

One of the challenges in ASR research nowadays is the recognition of casual, spontaneous speech (Benesty et al., 2008) such as human-to-human conversations. This type of speech is often not well articulated, lip smacks might occur, and may also contain disfluencies like "uh" or "well" as well as colloquial expressions like "she's out" instead of "she is going out" or "hey guys" instead of "hello, sir". Often, slurring pronunciations of certain phonemes, syllables or even larger segments of speech can be observed. The variability of temporal and spectral structures of spontaneous speech is larger in comparison to read speech (Benzeghiba et al., 2007). The *rate of speech* (ROS)[1] is a major variability that an ASR system for spontaneous speech has to cope with. The ROS is typically one variability that changes with the different speaking styles and affects the error rate of the ASR system: the higher the ROS the higher the error rate. Normal and low ROS does usually not affect performance. However, too slow, over-articulated speech does affect performance. Methods that try to reduce the effects of varying ROS were proposed for different domains of an ASR system, including codebook adaptation for the front-end, transition-probability adaptation of the acoustic models, and ROS-dependent language model adaptation (Siegler and Stern, 1995).

Inter-speaker Variabilities Inter-speaker variabilities refer to differences between speakers. Huang et al. (2001a) performed an analysis of estimated speaker-adaptation transformations with principal and independent component analysis (Bishop, 2006) with respect to different inter-speaker variabilities such as gender,

[1] No unique definition exists for the ROS. Besides "spoken words per minute" Siegler and Stern (1995) propose "phonemes per minute" as a more precise measure.

accent, and age. They showed that the first two principal components of inter-speaker variabilities correspond to the gender and to the accent, respectively. A more detailed analysis of the differences in the physiology of the speech apparatus of women and men was also part in the work of Boë et al. (2006). It was shown in that work, for example, that the pharyngeal cavity length of men reaches and even exceeds the length of the oral cavity during the first 20 years of life, while this is not the case for women. In average, the vocal tract length of females (about 14.5 cm) is smaller than that of men (about 17.5 cm), which leads to a natural variability already within the group of adult speakers. As the second principal component, regional and (even more) foreign accents degrade the performance of an ASR system (Huang et al., 2001a; Lawson et al., 2003). Native accents can be described by systematic shifts on the pronunciations and, thus, may be recognized by using native-accent specific acoustic models in parallel. In contrast to that, accentuated speech from nonnative speakers is much more difficult. Reasons for that are the influences of the native language and the proficiency in the nonnative language (Van Compernolle, 2001). Besides accent-specific acoustic models detailed pronunciations variants are commonly used for accented speech recognition (Lawson et al., 2003).

Age is another variability that has an impact on speech characteristics, and the recognition of children and elderly speech is another major issue in the field of ASR. Difficulties arise for children from their smaller physical size. This leads to higher positions of resonance frequencies of the vocal tract in comparison to adults. Furthermore, larger variations in formant locations and fundamental frequencies can be observed within the group of nonadult speakers. Also, the pronunciation of children may differ significantly from that of adults as well as the vocabulary and grammar. Wilpon and Jacobsen (1996) assume that the decrease in accuracy for elderly speakers can possibly be attributed to changes in the glottis area and the internal control loops of the human articulatory system. They presented results that suggest that ASR systems for speakers younger than 15 years and older than 70 years need special precautions to obtain an acceptable recognition performance.

Intra-speaker Variabilities On a sound wave level a human speaker is hardly able to exactly reproduce the same utterance twice. The complex interaction between all articulators and, furthermore, the conscious and unconscious neural signals lead to differences between the speech signals of a single speaker. Because of this the needed amount of templates for template-based recognition systems is very large in case of speaker-independent applications. *Dynamic time warping* (DTW) approaches together with vector quantization techniques, which were developed over the period from 1960 to 1980, need an increasing set of templates to recognize the variants of one and the same word. The introduction of the hidden Markov model methodology for ASR during the 1980ies provided a solution to this problem by statistically describing the distributions of the features. This allowed a compact parametrization that scales well with the amount of training data (Benesty et al., 2008).

On a phonetic level, *coarticulation* is another major source of variability in speech. Speech is the product of continuously moving articulators and the production of a certain sound is the product of a complex sequence of articulator movements. Speech production can be seen as articulator movements towards certain targets (Ladefoged, 2001). Coarticulation is the effect in which temporally neighbored targets influence the way the current sound is articulated. The extend to which coarticulation occurs depends on the one hand on the differences between the target positions of the intended sound and its previous and following sound. On the other hand it depends on the time interval between the targets. For example, in fast or spontaneous speech conversations many phonemes are often not fully realized.

The *Lombard effect* (Lombard, 1911) usually occurs in noisy environments. Generally, this is a condition where speech production is altered in an effort to communicate more effectively across a noisy environment (Hansen, 1996). The work of Hansen (1996) also gives an overview of how stress affects the speech characteristics. Stress causes changes in the characteristics of the glottal source, pitch, intensity, duration and spectral vocal tract characteristics. Automatic speech recognition experiments show a significant decrease in accuracy.

Variability also exists on the cognitive level of individual speakers. Here, the meaning of a word with the same pronunciation heavily depends on the communication context. The following exemplary utterance, taken from Huang et al. (2001b), demonstrates this:

> *Mr. Wright should write to Ms. Wright right away about his Ford or four door Honda.*

In this example the words "Wright", "write", and "right" all have the same phonetic realization, as well as, "Ford or" and "four door". The meaning not only depends on the grammatical role of the word, but also on *pragmatical knowledge*. This is knowledge about the context of the utterance, about the communication partners, their intentions, common sense, etc.

1.3. Overview

Methods for extracting robust parametric representations from the speech signals as input to ASR systems are in the focus of this work. Different methods have been investigated for increasing the robustness of speech recognizers against some variabilities as described above. The focus of the proposed methods lies on the inter-speaker variability that originates from different vocal tract lengths. Another emphasis is put on methods that try to increase the robustness against environmental variabilities like additive noise. The main contributions of this thesis can be summarized as following:

- A major focus is put on different feature extraction methods that make use of invariance transforms. Feature-extraction methods are presented that make use of two translation-invariant methods that were originally proposed within the field of image analysis and recognition. Furthermore, a general approach for the construction of invariant features known as "invariant integration" is adapted to the field of ASR and a feature extraction method that is based on this approach is described.
- The use of an auditory model that originally has been proposed to explain certain psycho-acoustical observations is combined with one of the proposed invariant feature extraction methods. It is shown that the features based on this model provide supplementary information to features that are based on a commonly used filter bank and further improve the performance of the recognizer under noisy conditions.
- Two methods for the estimation of the spectral effects due to changes of the vocal tract length are presented in Chapter 6. The first method is data-driven and allows for the computation of a deformation that relates two time-frequency representations to each other. It is shown that the resulting deformations can be used to enhance a commonly used method for vocal tract length normalization. The second method is model-driven and estimates the spectral effects with the help of a wave-reflection model of the vocal tract.

The structure of this thesis is as following: Chapter 1 gave a brief introduction about the human speech production and perception. Furthermore, different variabilities that a recognition system might have to handle to achieve a high performance were presented. Chapter 2 introduces basic principles and methods, which will consistently be referred to in the succeeding chapters. The architecture of the ASR system that is used throughout this work is described in Chapter 3. The different invariant feature types and methods for their application are described in detail in Chapter 4. Methods that increase the noise robustness of an ASR system during the feature extraction are presented in Chapter 5. The methods for estimating the spectral effects due to different vocal tract lengths are described in Chapter 6. A summary of the methods presented in this thesis together with a discussion and an outlook is given in the final Chapter 7. Information about the used data sets, the mathematical notation, and additional details of experiments are provided in the Appendix.

2

Basic Concepts

The previous section schematically described the speech production and recognition process. More formally, a speech recognition problem may be formulated as following: A speaker decides on a source word sequence $\boldsymbol{W} = (w_1, w_2, \dots, w_P)$, generates the corresponding speech signal, and transmits the signal through a (possibly noisy) communication channel. The receiver processes the obtained signal and decodes the acoustic signal into an estimated word sequence $\widehat{\boldsymbol{W}} = (\widehat{w}_1, \widehat{w}_2, \dots, \widehat{w}_{\widehat{P}})$. Ideally, the source word sequence $\boldsymbol{W}$ and the estimated word sequence $\widehat{\boldsymbol{W}}$ are the same. In the following, an overview of the concepts and basic methods that are involved within a modern speech recognition system is given.

2.1. Basic Architecture of a Speech Recognition System

Currently, the principal structure of a *large-vocabulary continuous speech recognition* (LVCSR) system follows a mathematically based, statistical modeling framework. *Hidden Markov models* (HMM, Rabiner and Juang (1986)) are the base of this approach and until today they are a central component of all speech recognition systems. As is described in more detail in the following, HMMs provide the possibility to statistically combine linguistic structure with the intrinsic variability of

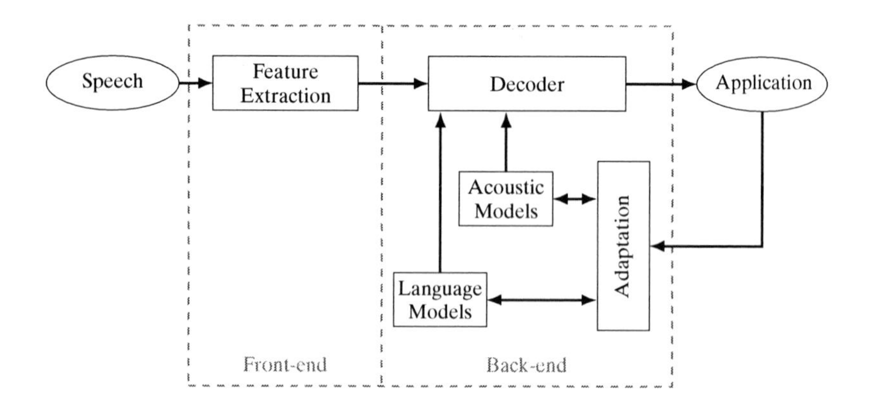

Figure 2.1.: Basic architecture of a speech recognition system. Adapted from Huang et al. (2001b); Benesty et al. (2008).

speech signals. The linguistic structure of speech can be modeled with a Markov chain of states, while the acoustic features of speech are described with a set of probability density distributions (Benesty et al., 2008). The basic architecture of a LVCSR system is shown in Figure 2.1. The *front-end* of a speech recognition systems extracts a parametric representation from the audio signal and processes this signal such, that an appropriate representation of the contained linguistic information is obtained. To accomplish this, the received signal is converted into short segments of equal length. In the context of ASR the segments are typically called "*frames*". A parametric representation is computed for each frame, which yields a sequence $\boldsymbol{Y}_{1:T} \in \mathbb{R}^{d \times T}$ of length T with feature vectors $\boldsymbol{y} \in \mathbb{R}^d$,

$$\boldsymbol{Y}_{1:T} = \begin{bmatrix} \boldsymbol{y}_1 & \boldsymbol{y}_2 & \dots & \boldsymbol{y}_T \end{bmatrix}. \tag{2.1}$$

For ease of notation, the subscript of $\boldsymbol{Y}_{1:T}$ is omitted if appropriate in the following. The decoder as part of the *back-end* aims to find the sequence of words $\widehat{\boldsymbol{W}}$ that is most likely to have generated the observed sequence of features $\boldsymbol{Y}$,

$$\widehat{\boldsymbol{W}} = \arg\max_{\widetilde{\boldsymbol{W}}} \left[p(\widetilde{\boldsymbol{W}} | \boldsymbol{Y}) \right]. \tag{2.2}$$

Bayes' rule is used to restate the problem in Equation (2.2) and to make the problem more feasible,

$$\widehat{\boldsymbol{W}} = \arg\max_{\widetilde{\boldsymbol{W}}} \left[p(\boldsymbol{Y} \mid \widetilde{\boldsymbol{W}}) \cdot P(\widetilde{\boldsymbol{W}}) \right]. \tag{2.3}$$

The likelihood $p(\boldsymbol{Y}|\widetilde{\boldsymbol{W}})$ in Equation (2.3) is computed with *acoustic models*, which contain information about phonetics, microphone and environmental variability, as well as variabilities among speakers. The prior $P(\widetilde{\boldsymbol{W}})$ is computed with a *language model*. Language models contain information about possible words, co-occurrence likelihoods, and semantics and functions of the ASR system. A target application receives the decoding result. Furthermore, it provides data that can be used for adapting the acoustic and/or the language models to increase the systems robustness against different variabilities. In the remaining of this chapter, each of the mentioned components is described in more detail.

2.2. Feature Extraction Basics

Given a speech waveform that has been sampled into a digital signal, the feature extraction stage of an ASR system is supposed to transform the speech signal $x(m)$ into a suitable representation for the subsequent stages of the system. The representation should only contain the information that is necessary for the discrimination between the classes of interest. Furthermore, their distribution in the feature space should fit to the assumptions made by the acoustical models in the back-end of the ASR system. Typically, feature vectors $\boldsymbol{y}_t$ for speech recognition are computed for short segments of the input signal. The segments are commonly referred to as *frames* and have a length τ_l of 20 to 25 ms, $\tau_l \in \mathbb{R}^+$. It is assumed that the characteristics of the speech signal remain constant over this time interval. A typical frame shift τ_s is 10 ms, $\tau_s \in \mathbb{R}^+$. Commonly used feature extraction methods imitate physiological and psychoacoustical findings about the human auditory system up to a certain degree. Examples of often used feature extraction methods are *mel-frequency cepstral coefficients* (MFCC) (Davis and Mermelstein, 1980) and *perceptual linear prediction* (PLP) coefficients (Hermansky, 1990).

The next section gives an overview of commonly used approaches for a time-frequency analysis of speech signals in the context of ASR. First, a brief introduction of the short-time Fourier transform as classical tool for a time-frequency analysis is given. Afterwards, auditory filter banks and the methods for the computation of MFCC and PLP features are described. The use of the frame-wise computed logarithmized energy of the time signal has proven to generally increase the performance of ASR systems and is described in Section 2.2.4. Different ways for the incorporation of contextual information into the features are described afterwards and methods for enhancing the robustness of the features to noise are described in Section 2.2.6.

2.2.1. Time-Frequency Analysis

In order to find a parametric representation of the speech signal that allows for the recognition of the linguistic content, a central idea within the front-ends of automatic speech recognition systems is to imitate physiological and psychoacoustical findings up to a certain degree (see also Section 1.1). A key observation is that the cochlea operates as a short-time Fourier analyzer, which separates acoustic signals into their frequency components (Moore, 1995, pp. 75). Within the ASR system this functionality is realized with a filter bank that yields a time-frequency representation of the input speech signal. Starting with uniformly distributed bandpass filters (Rabiner and Juang, 1993, pp. 77) with pure signal-processing backgrounds a few decades ago, various more sophisticated approaches have been proposed that better explain psychoacoustic findings. An overview of the principle ideas of these models can be found in the work of Walters (2011). In the following, filter banks that were considered within this work and that incorporate psychoacoustic findings up to different levels of detail are briefly described next. A sophisticated model for the human auditory perception referred to as "auditory image model" is investigated for a possible application in ASR in combination with invariant feature types in Section 5.3.

Automatic speech recognition processes discrete-time (speech) signals. A classical method for a time-frequency analysis is the *short-time Fourier transform* (STFT) (see, for example, Mertins, 1994; Oppenheim and Schafer, 1999). The idea of the STFT is to obtain a time-dependent representation of spectral components by first multiplying the discrete-time signal $x(m)$ with an analysis window $w(m)$ and, second, by computing the Fourier transform of the windowed signal. Formally, the *short-time discrete-time Fourier transform* (stDTFT) can be written as

$$S(n, \omega) = \sum_{m=-\infty}^{\infty} x(m) w(m-n) e^{-j\omega m} \tag{2.4}$$

In this formulation, ω is a continuous parameter. Equation (2.4) can be interpreted as the convolution of the complex sequence $x(m)e^{-j\omega m}$ with the (real valued) window sequence $w(m)$. Commonly used window types are the Hamming and the Hann window. For a given sampling rate s_r, the number of samples $\tau_l^d \in \mathbb{N}^+$ or $\tau_s^d \in \mathbb{N}^+$, which correspond to a length of τ_l or τ_s, respectively, can be computed by

$$\tau_l^d = \lfloor s_r \cdot \tau_l \rfloor, \tag{2.5}$$

$$\tau_s^d = \lfloor s_r \cdot \tau_s \rfloor. \tag{2.6}$$

In practice, the length of the analysis window is set to τ_l^d. The Hamming and the Hann window (together with additional often used windows) are shown in

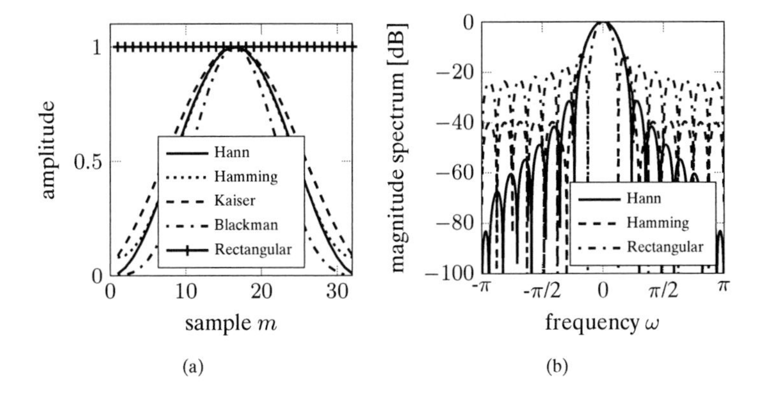

Figure 2.2.: (a) Plots of the rectangular, Hann, Hamming, Kaiser, and Blackman window functions with window length $\tau_l^d = 32$, (b) magnitude spectrum of Hann, Hamming, and rectangular windows with window length $\tau_l^d = 16$.

Figure 2.2 (a). The *Hamming window* (Harris, 1978) is defined as

$$w(m) = \begin{cases} 0.54 - 0.46\cos(\frac{2\pi m}{\tau_l^d - 1}), & m = 0, 1, \ldots, \tau_l^d - 1, \\ 0, & \text{otherwise,} \end{cases} \tag{2.7}$$

the *Hann window* is defined as

$$w(m) = 0.5\left(1 - \cos\left(\frac{2\pi m}{\tau_l^d - 1}\right)\right), \tag{2.8}$$

and the *rectangular window* is defined as

$$w(m) = \begin{cases} 1, & m = 0, 1, \ldots, \tau_l^d - 1, \\ 0, & \text{otherwise.} \end{cases} \tag{2.9}$$

A trade-off between preserving the temporal and the spectral characteristics of the signal is made by the choice of the window function. The rectangular window does not apply any changes by weighting the temporal signal $x(m)$. It can be seen in Figure 2.2 (b) that the height of the side lobes is about -20 dB with respect to the main lobe. Smoother windows like the Hann or the Hamming window generally have a wider main lobe, but the attenuation in the side lobes is much larger compared to the rectangular window (see also Figure 2.2 (b)). Because of this characteristics, smoother windows are usually preferred in comparison to the rectangular window. Usually, the reconstruction of the time signal from (processed) spectral values is not necessary for the feature extraction in ASR. Thus, the necessary conditions for a reconstruction of the time signal are not further addressed here. Figure 2.3 (a) illustrates this approach.

By considering only the discrete frequencies ω_k of the stDTFT, where

$$\omega_k = 2\pi k/K, \qquad k = 0, \ldots, K-1, \tag{2.10}$$

the complex spectral value $s_n^c(k)$ of the *short-time discrete Fourier transform* (stDFT) at time index n and for frequency bin k can be calculated as

$$s_n^c(k) = \sum_{m=-\infty}^{\infty} x(m)w(m-n)W_K^{km}, \qquad W_K = e^{-j2\pi/K}. \tag{2.11}$$

By choosing a frame length τ_l^d for the analysis window, and by introducing the frame shift τ_s^d into the time-frequency analysis as given by Equation (2.11), a stDFT with a given frame length and shift can be computed by

$$s_n^c(k) = \sum_{m=n\tau_s^d-\tau_l^d+1}^{n\tau_s^d} x(m)w(m-n\tau_s^d)W_K^{km}, \qquad W_K = e^{-j2\pi/K}. \tag{2.12}$$

In the context of this work, the notions of "short-time DFT" and "short-time Fourier transform" are used interchangeably. In practice, the *fast Fourier transform* (FFT, Golub and Van Loan (1996)) can be used for an efficient computation of Equation (2.11).

As is described in more detail in the following sections, feature extraction methods for speech recognition tasks usually rely on the magnitude of the spectral values $s_n^c(k)$, which is called *spectrogram*. Thus, the notion of "spectral values" in the rest of this work refers to the magnitude of the spectral values $s_n^c(k)$, if not stated otherwise, and is defined as

$$s_n(k) := |s_n^c(k)|. \tag{2.13}$$

The spectrogram of the speech signal as shown in Figure 2.3 (a) is illustrated in Figure 2.3 (b).

The human auditory system is able to separate the components of a sound up to a certain degree. The frequency resolution and perception of the auditory system, as well as the loudness perception has been the center of many studies in the past decades (Fletcher, 1953; Moore, 1995, and references therein). Even though the cochlea is not completely understood until today, it is commonly assumed that it has

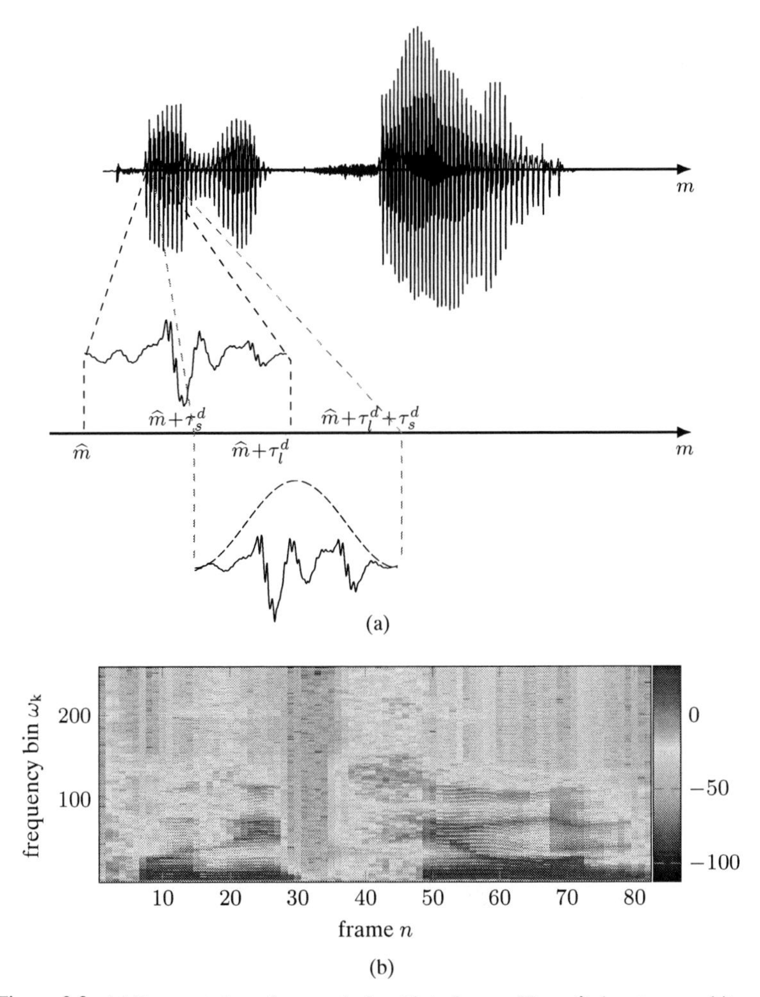

Figure 2.3.: (a) Segmentation of a speech signal into frames. Here, $\widehat{m}$ denotes an arbitrary sample number, τ_l^d denotes the frame length in number of samples and τ_s^d denotes the frame shift in number of samples. A Hamming analysis window was used as indicated by the dashed curve at the second signal segment. For each frame a feature (observation) vector is computed, which yields a sequential parametric representation of the speech signal. (b) Spectrogram of the signal with $\tau_l = 25$ ms and $\tau_s = 10$ ms and a Hamming window.

a central role in the frequency analysis process. Different types of filter banks have been proposed that explain the observations of psychoacoustic experiments about the frequency analysis. The mel, gammatone, and gammachirp filter banks are three examples that are also used within the field of ASR. In the following, these are described in more detail.

Mel Filter Bank The *mel filter bank* makes use of findings from psychoacoustic experiments that have lead to the *mel scale*. The mel scale relates the perceived pitch to frequency. With its origins in the work of Stevens and Volkmann (1936), the mel scale is constructed with the view to explain the perception of pure tones. It is often described as being linear up to 1000 Hz and the upper region being logarithmic (O'Shaughnessy, 1999). However, it has been shown by Umesh et al. (1999b) that there is insufficient evidence for this explicit model. Until today, there are different equations that fit the experimentally obtained mel scale. For mappings from the frequency to the mel domain we use the equation that can be found, for example, in the works of O'Shaughnessy (1999); Shannon and Paliwal (2003); Young et al. (2009):

$$\mathrm{mel}(f) = 2595 \cdot \log_{10}\left(1 + \frac{f}{700}\right), \tag{2.14}$$

with f in Hz. Equivalently, the mapping $\mathrm{mel}^{-1}(g)$ from the mel domain to the frequency domain is given by $\mathrm{mel}^{-1}(g)$,

$$\mathrm{mel}^{-1}(g) = 700 \cdot 10^{\frac{g}{2595}} - 1. \tag{2.15}$$

Perceptually motivated filter banks like the mel filter bank commonly locate the filter center frequencies evenly located on auditory scales. Because these scales need to be used in a parametrized form in later chapters, they are introduced in a more general way in the following: Generally, with given minimum and maximum frequencies f_{min} and f_{max}, respectively, K evenly spaced center frequencies $\boldsymbol{f}_c = (f_c(1), f_c(2), \ldots, f_c(K))$ with respect to a given scale $\xi(f)$ can be computed by first determining the scale-dependent step size Δ_ξ,

$$\Delta_\xi = \frac{1}{K-1}\left[\xi(f_{\mathrm{max}}) - \xi(f_{\mathrm{min}})\right]. \tag{2.16}$$

Second, the center frequencies $f_c(k)$ are given by

$$f_c(k) = \xi^{-1}\left(\xi(f_{\mathrm{min}}) + (k-1)\Delta_\xi\right). \tag{2.17}$$

In case of a mel filter bank it is $\xi(f) = \mathrm{mel}(f)$. The filters of a mel filter bank are usually triangular shaped. Their support is commonly chosen such that it covers the frequency range from the center of the previous triangular filter to the center of the next triangular filter. The filters are used to compute weighted sums of the spectral values from a previous STFT step: We assume to have K_1 spectral values $\boldsymbol{s}_n^{\text{lin}} \in \mathbb{R}^{K_1}$ for frame n,

$$\boldsymbol{s}_n^{\text{lin}} = \left(s_n^{\text{lin}}(1), s_n^{\text{lin}}(2), \ldots, s_n^{\text{lin}}(K_1)\right)^{\mathrm{T}}, \tag{2.18}$$

as result of a STFT. Let a mel filter bank have K_2 subbands in the following. Now, we want to compute the filter weights $\boldsymbol{h}_l \in \mathbb{R}^{K_1}$,

$$\boldsymbol{h}_l = (h_l(1), h_l(2), \ldots, h_l(K_1))^{\mathrm{T}}, \qquad 1 \le l \le K_2, \tag{2.19}$$

for each subband l, $1 \le l \le K_2$, $K_1 > K_2$, of a mel filter bank. Let

$$\boldsymbol{f}_c^{\text{mel}} = \left(f_c^{\text{mel}}(1), f_c^{\text{mel}}(2), \ldots, f_c^{\text{mel}}(K_2)\right) \tag{2.20}$$

denote the center frequencies of the mel filter bank that have been computed according to Equations (2.14) to (2.17). Furthermore, let

$$f_c^{\text{DFT}}(k) = \frac{(k-1)}{2K_1} \cdot s_r, \qquad k = 0, 1, \ldots, K_1 - 1, \tag{2.21}$$

denote the center frequencies of the STFT bins, where s_r is the sampling rate. Then, for each subband l of the mel filter bank, the corresponding filter weights $h_l(i)$ are given by

$$h_l(i) = \begin{cases} \frac{f_c^{\text{DFT}}(i) - f_c^{\text{mel}}(l-1)}{f_c^{\text{mel}}(l) - f_c^{\text{mel}}(l-1)}, & f_c^{\text{DFT}}(i) > f_c^{\text{mel}}(l-1) \text{ and } f_c^{\text{DFT}}(i) \le f_c^{\text{mel}}(l), \\ \frac{f_c^{\text{mel}}(l+1) - f_c^{\text{DFT}}(i)}{f_c^{\text{mel}}(l+1) - f_c^{\text{mel}}(l)}, & f_c^{\text{DFT}}(i) > f_c^{\text{mel}}(l) \text{ and } f_c^{\text{DFT}}(i) < f_c^{\text{mel}}(l+1), \\ 0, & \text{otherwise.} \end{cases} \tag{2.22}$$

Sometimes, the triangular filters are area normalized,

$$h_l'(i) = \frac{1}{\sum_{i'=1}^{K_1} h_l(i')} \cdot h_l(i), \quad 1 \le l \le K_2. \tag{2.23}$$

By summarizing the filter weights within a matrix $\boldsymbol{H}$ as

$$\boldsymbol{H} = \begin{bmatrix} \boldsymbol{h}_1 & \boldsymbol{h}_2 & \cdots & \boldsymbol{h}_{K_2} \end{bmatrix}, \tag{2.24}$$

the mel weighted spectral values $\boldsymbol{s}_n^{\text{mel}} \in \mathbb{R}^{K_2}$ can be computed as

$$\boldsymbol{s}_n^{\text{mel}} = \boldsymbol{H}^{\mathrm{T}} \boldsymbol{s}_n^{\text{lin}}. \tag{2.25}$$

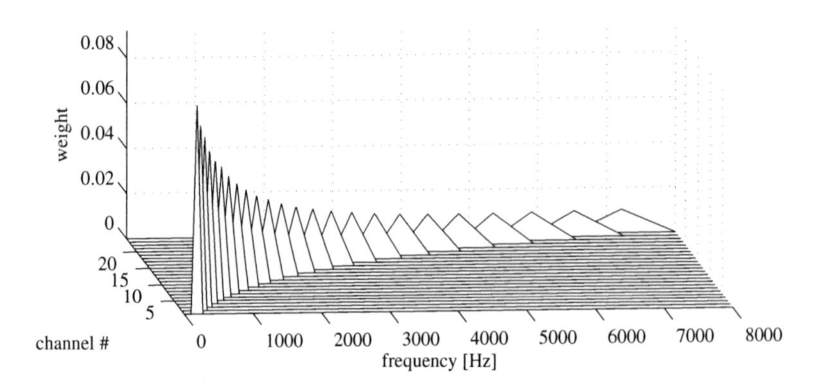

Figure 2.4.: Mel filter bank weights $\boldsymbol{H}$ with 24 triangular-shaped filters, here with the area normalized to unity, $f_{\text{min}} = 40$ Hz, $f_{\text{max}} = 8000$ Hz.

If it is clear from the context, the superscripts of $\boldsymbol{s}_n^{\text{lin}}$ or $\boldsymbol{s}_n^{\text{mel}}$ are omitted in the following. Figure 2.4 shows exemplary triangular filters whose center frequencies are equally spaced along the mel scale. The mel filter bank is a central component for the computation of mel frequency cepstral coefficients as features for ASR. These are explained in detail in Section 2.2.2.

Gammatone Filter Bank A more accurate filter shape compared to the triangular shapes used in the mel filter bank is used in case of the gammatone filter bank. This approximation for the time-frequency analysis stage within the inner ear makes use of *gammatone filters*. These filters are defined in the time domain as (Patterson et al., 1992):

$$g_t(t) = t^{n-1} \exp(-2\pi bt) \cos(2\pi f_c t + \phi), \qquad t \geq 0, \tag{2.26}$$

where b determines the bandwidth of the filter, n is the order, f_c determines the center frequency, and ϕ is the phase of the impulse response. A measure for the auditory frequency resolution of the human ear is described by the *equivalent rectangular bandwidth* (ERB). Based on notched-noise experiments, Glasberg and Moore (1990) proposed an equation that describes the ERB as a function of center frequency f,

$$\text{ERB}(f) = 24.7 \cdot (4.37f/1000 + 1)\,. \tag{2.27}$$

With $n = 4$ Patterson et al. (1992) show that the bandwidth parameter $b(f)$ of the gammatone filter from Equation (2.26) can be chosen as

$$b(f) = 1.019 \cdot \mathrm{ERB}(f). \tag{2.28}$$

The ERB scale ERBS(f) is defined as the number of ERBs below the given frequency f in Hz (Moore and Glasberg, 1996),

$$\mathrm{ERBS}(f) = 21.4 \cdot \log_{10}(0.00437f + 1). \tag{2.29}$$

With given minimum and maximum frequencies f_{min} and f_{max}, respectively, and with a given number of channels K, Slaney (1993) derived an expression for filter center frequencies f_{c_k} that are located equally spaced on the ERB scale,

$$f_c^{\mathrm{ERB}}(k) = -24.7\zeta + \exp\left(k \cdot \left(-\log\left(f_{\mathrm{max}} + 24.7\zeta\right) + \log\left(f_{\mathrm{min}} + 24.7\zeta\right)\right)/K\right) \cdot \left(f_{\mathrm{max}} + 24.7\zeta\right), \qquad k = 1, \ldots, K. \tag{2.30}$$

Here, $\zeta = 9.26449$ (Glasberg and Moore, 1990). The gammatone filter bank used in this work is based on the implementation by Ellis (2009), which, in turn, is based on the technical report from Slaney (1993). Similar to the mel filter bank approach, a weighting matrix $\boldsymbol{H}$ is computed that contains a discrete approximation of the magnitudes of the transfer function of the gammatone filters as given in Equation (2.26). Figure 2.5 shows the log-magnitudes of a gammatone filter bank with 24 channels and a frequency range from 40 to 8000 Hz.

Dynamic-Compressive Gammachirp Filter Bank Originally, the gammatone function was used to characterize impulse response data from auditory fibers in the cat. As argued by Irino and Patterson (1997), the use of the gammatone filter is limited by demonstrations that showed that the skirt of the auditory filter broadens below its center frequency with increasing stimulus level and sharpens a little above its center frequency. Irino (1996) derived the *gammachirp function* whose real part of the analytic gammachirp function was introduced as the *gammachirp auditory filter*. The impulse response of the gammachirp auditory filter is defined as

$$g_c(t) = t^{n-1} \exp\left(-2\pi b \, \mathrm{ERB}(f_r)t\right) \cos\left(2\pi f_r t + c \ln t + \phi\right), \qquad t > 0, \tag{2.31}$$

where, similar to Equation (2.26), b determines the bandwidth of the filter, n is the order, and ϕ is the phase of the impulse response. Following the notation of Irino and Patterson (1997), f_r is used instead of f_c here, because the peak frequency of the amplitude spectrum varies with c and, to a lesser extend, with b and n. In

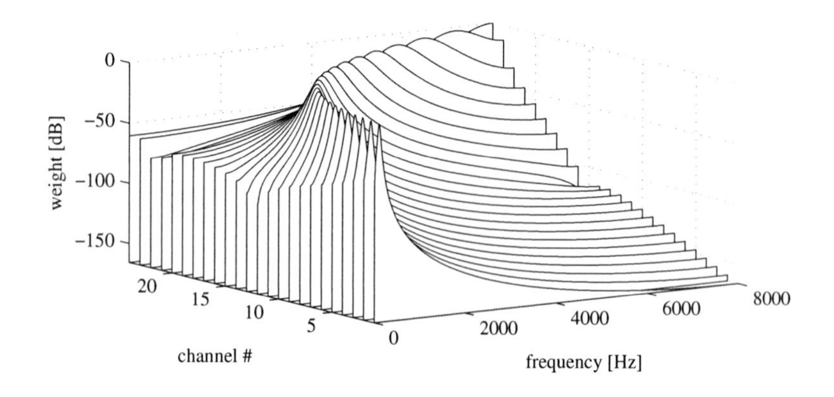

Figure 2.5.: Filter weights of a 24-channel gammatone filter bank.

addition to the definition of a gammatone filter in Equation (2.26), the cosine term contains $c \ln t$, where c is an additional parameter. By fitting gammachirp filters to measured psychophysical data, Irino and Patterson (1997) showed that the gammachirp filter is a better approximation that data than gammatone filters. Figure 2.6 shows the auditory filter shapes for different sound pressure levels.

The peak frequency in that figure is normalized to 2000 Hz to clarify the level-dependent asymmetry of the filter shape. The asymmetric shape for different probe levels is clearly visible. However, the gammachirp function was derived from notched-noise masking data that limited the center frequencies to the region of about 2 kHz. The publication of extended masking data led to the development of the *compressive gammachirp* (cGC) auditory filter as presented by Patterson et al. (2003). It consists of a passive gammachirp filter and an asymmetric function that shifts in frequency with stimulus level. A nonlinear auditory filter bank with analysis/synthesis capabilities was presented by Irino and Patterson (2006) and is known as *dynamic-compressive gammachirp* (dcGC) filterbank. It is based on the cGC auditory filter and uses a level control circuit, which dynamically adapts the compression parameter of the cGC filters in response to the input signal. A Matlab implementation of the dcGC filter bank is available upon request from Irino. In this work, a Matlab implementation was used that has been received from Irino in April, 2008.

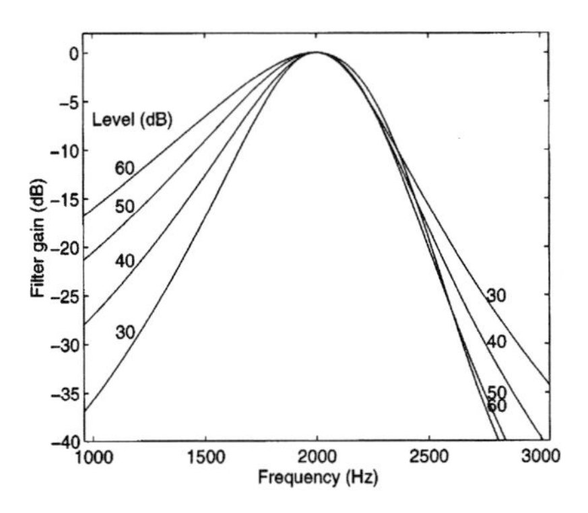

Figure 2.6.: Gammachirp auditory filter shape as function of probe level (30-60 dB sound pressure level) with center frequency normalized to 2000 Hz. The figure is taken from the work of Irino and Patterson (1997).

2.2.2. Mel Frequency Cepstral Coefficients

Mel frequency cepstral coefficients (MFCC) are currently one of the simplest, but most widely used features for ASR. The motivation for the use of MFCC features for ASR is given by considering the source-filter model that is described in Section 1.1: The speech signal $x(m)$ can be seen as being the result of a source signal passed through a vocal tract filter. By modeling this as a linear, time-invariant system, the speech signal is generated by convolving the source signals with the impulse response of the vocal tract filter. A convolution in the time-domain corresponds to a multiplication in the spectral domain. Taking the logarithm of the magnitudes of the spectral values turns the multiplicative relation into an additive relation. Within this domain, it is assumed that the source signal leads to more rapid changes along the time axis than the spectral characteristics of the vocal tract. In case of cepstral coefficients, a DCT is used to separate the low-frequency components, which are assumed to correspond to the vocal tract characteristics, from the rest of the signal. The notions "cepstrum" and "cepstral" originate from the idea of performing a spectral analysis within the spectral domain. A schematic overview for the computation of MFCCs is shown in Figure 2.7. A typical first step for the computation of MFCC features is the use of a *pre-emphasis filter*

$$H_P(z) = 1 - \mu z^{-1}, \tag{2.32}$$

where a typical choice for μ is in the range of 0.95 and 0.99. This high-pass filter tries to eliminate the larynx and lip radiation characteristics, such that a succeed-

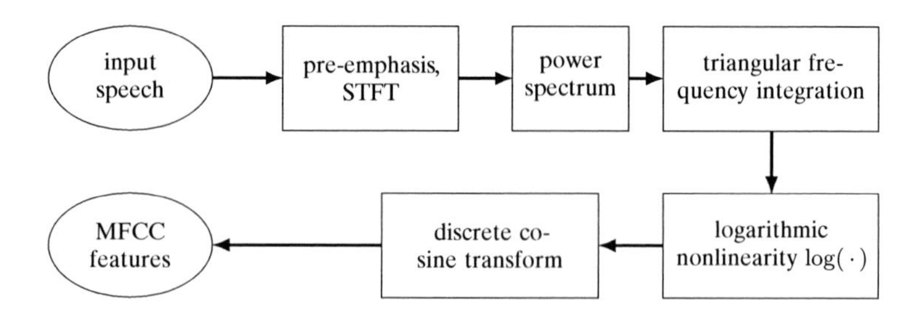

Figure 2.7.: Feature extraction stages for the computation of mel frequency cepstral coefficients (MFCC).

ing TF analysis describes the parameters from the vocal tract only (Deller et al., 1993, p. 329). As described in Section 2.2.1 the input speech signal is then framed into segments with a length of 20-25 ms and weighted by a Hamming analysis window. For each segment n the magnitude of the stDFT (see Equation (2.11)) is computed. The result is passed to 20 and 30 triangular-shaped filters with centers located equally spaced on the mel scale (see also Figure 2.4) to compute mel weighted spectral values $\tilde{s}_k(n)$. As described in more detail in the previous section the mel scale is defined as being approximately linear in frequency up to 1 kHz and logarithmic at higher frequencies. The well-known MFCCs are computed by taking the logarithm of the mel weighted spectral values and by transforming these with a *discrete cosine transform* (DCT),

$$y_n(k) = \sum_{k'=0}^{K-1} \log\left(\tilde{s}_n(k')\right) \cos\left[\frac{k\pi}{K}\left(k' + \frac{1}{2}\right)\right], \tag{2.33}$$

The DCT step in Equation (2.33) can efficiently be computed by mirroring the real-values spectral values and using the FFT. Typically, the first 13 DCT coefficients are used as features. Often, ASR systems replace the first coefficient representing the mean value of the log-spectral values by the logarithmized energy of the signal segment (see also Section 2.2.4).

Davis and Mermelstein (1980) were among the first to use the mel scale for computing cepstral coefficient based features for automatic speech recognition and empirically showed that the use of the mel scale instead of a linear frequency scale improves the recognition accuracy.

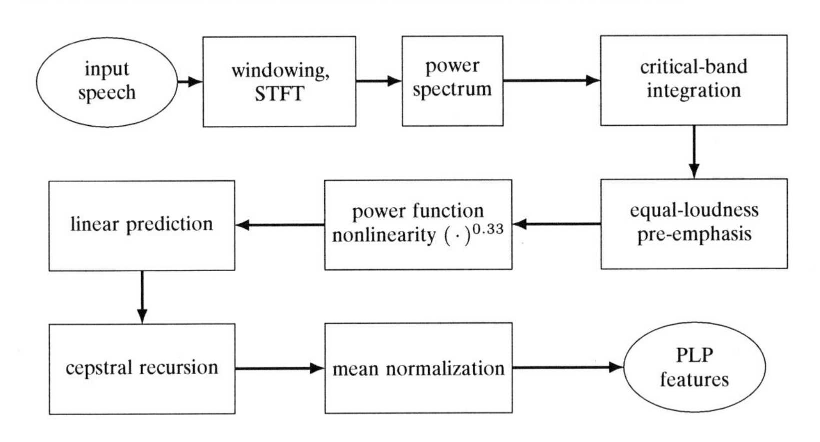

Figure 2.8.: Feature extraction stages for the computation of perceptual linear prediction (PLP) features.

2.2.3. Perceptional Linear Prediction Coefficients

Perceptual linear prediction (PLP) coefficients are another kind of commonly used features in ASR systems. The technique was proposed by Hermansky (1990) and simulates, in comparison to MFCC features, more findings of psychoacoustic experiments by engineering approximations. For the computation an auditory-like spectrum of the speech signal is computed first. Then, the resulting spectrum is parametrized with *linear prediction* (LP) (Makhoul, 1975), which is a method to identify the parameters of an autoregressive all-pole model. In the best case, only the characteristics of the vocal tract are captured by the model. An appropriate model order has to be chosen and determines the amount of details that can be described by the model.

The computation of the PLP coefficients is illustrated in Figure 2.8. After computing the power-spectrum of windowed segments of the speech signal (as described in the section above) a critical band analysis is done. In contrast to the MFCCs, the simulated critical-band masking curves have a trapezoidal shape. The output of the critical band integration is then passed to a processing block where the nonequal loudness sensitivity of human hearing at different frequencies is simulated by a corresponding weighting of the critical bands. Afterwards, a cubic-root amplitude compression is applied, which simulates the power-law of hearing (Stevens, 1957). The output of this stage is then parametrized by an auto-regressive model, whose parameters are determined efficiently with linear prediction (Makhoul, 1975). Ideally, formants are well approximated by the model, while finer structures (for example, those from the glottis or noise) are smoothed out. With the model order specifying the amount of details that can be described, the choice of order generally affects the recognition accuracy. Reports of "optimal" model orders for ASR are ambiguous and orders between five and 19 can be found in literature (for example, Hermansky, 1990; Sainath et al., 2009). Finally, the LP coefficients can be transformed into cepstral domain (Rabiner and Juang, 1993, p. 112ff) and/or mean normalized to increase their robustness to channel variabilities (see also Section 2.2.6).

2.2.4. Short-Time Energy

In case of features based on cepstral coefficents, the DC-coefficient is usually replaced by the logarithmized short-time energy (log-energy) as feature. In this work, the log-energy is computed as described by Young et al. (2009). Basically, it is the logarithmized signal energy within every considered frame. For speech samples $x(m)$, $m = 1, \ldots, \tau_l^d$ of a single frame with length τ_l^d the energy feature E is basically computed as

$$E = \log_{10} \sum_{m=1}^{\tau_l^d} x(m)^2. \tag{2.34}$$

The log-energy is usually normalized to a range of about 50 dB.

2.2.5. Inclusion of Temporal Context Information

To incorporate information about temporal context into the extracted features, two approaches are common for ASR systems nowadays. One approach is to adjoin successive feature vectors to a "super-vector" and subsequently reduce the dimensionality of this vector by a transformation as result of, for example, principal component analysis or linear discriminant analysis (see Section 2.3). The other approach is to supplement the feature vectors with *dynamic features*. These are usually represented as approximated time differentials. With $y_n(k)$ being the k-th feature component at time-instance n and $2M + 1$ being the number of frames to be considered for the linear regression, Rabiner and Juang (1993, p.196ff) derive approximations for the first and second order temporal derivatives,

$$\left. \frac{\partial y_{n+\tau}(k)}{\partial \tau} \right|_{\tau=0} \simeq \frac{1}{T_M} \sum_{\tau=-M}^{M} \tau y_{n+\tau}(k) \tag{2.35}$$

and

$$\left.\frac{\partial^2 y_{n+\tau}(k)}{\partial \tau^2}\right|_{\tau=0} \simeq \frac{2\left\{T_M\left[\sum_{\tau=-M}^{M} y_{n+\tau}(k)\right]-(2M+1)\left[\sum_{\tau=-M}^{M} \tau^2 y_{n+\tau}(k)\right]\right\}}{T_M^2-(2M+1)\left[\sum_{\tau=-M}^{M} t^4\right]}, \tag{2.36}$$

where

$$T_M = \sum_{\tau=-M}^{M} \tau^2. \tag{2.37}$$

In case of the approximation for the second order derivative, M is usually chosen to be larger for the approximation because of its higher sensitivity compared to the first order derivative. Common values are $M = 3$ and $M = 5$ for the approximated first and the second order derivatives, respectively. Often, the dynamic features are also referred to as "delta and delta-delta" ($\Delta + \Delta\Delta$) features. Both terms are used interchangeably in this work.

2.2.6. Enhancement of Noise Robustness

As explained in the introduction, automatic speech recognition systems have to deal with different kinds of variabilities with background noise being one of them. Besides feature representations that try to be immune to noise, for example, RASTA-PLP (Hermansky and Morgan, 1994), many methods have been proposed to compensate for the acoustic mismatch between training and testing data due to noise. An overview of these methods can be found in the book of Gales and Young (2008). Generally, these methods are either speech feature or model enhancement techniques, and some methods can be seen as hybrid approaches. Cepstral mean normalization, *stereo piecewise linear compensation for environment* (SPLICE) (Deng et al., 2000), and vector Taylor-series (VTS) expansion (Moreno et al., 1996) are exemplary methods for feature enhancement methods. Another feature enhancement method that is based on maximizing the sharpness of the power distribution and on power flooring was recently proposed (Kim and Stern, 2010a) and yields so-called *power-normalized cepstral coefficients* (PNCC). Generally, all these methods try to remove the effects of noise from the feature vectors to reduce the mismatch between training and testing data.

Parallel model combination (PMC) (Gales, 1995) is one example for the group of model enhancement techniques, where the parameters of the clean acoustic models are adapted such that they approximate the model parameters of training with corrupted speech. Since this work focuses on feature-based enhancement approaches, methods of this group are not further considered in the following. An advantage of feature enhancement methods are the smaller computational costs compared to the model adaptation techniques. In the following, the feature enhancement methods that were used in this work are explained in more detail.

Mean and Variance Normalization Due to its efficiency and effectiveness, one of the most commonly used feature normalization methods for increasing the noise robustness within ASR systems is mean normalization. This procedure is motivated by considering a clean speech signal $x(m)$ that is passed though a linear, time-invariant system, whose channel characteristic is given by the transfer function $H(n,\omega)$. With $S(n,\omega)$ being the stDTFT of the clean speech signal, the output $\boldsymbol{Y}(n,\omega)$ of the system is given by

$$\boldsymbol{Y}(n,\omega) = S(n,\omega) \cdot H(n,\omega). \tag{2.38}$$

Now, by computing the logarithm of the magnitudes of the spectrum,

$$\log_{10} \boldsymbol{Y}(n,\omega) = \log_{10} S(n,\omega) + \log_{10} H(n,\omega), \tag{2.39}$$

it can be seen that the multiplicative effect of the filter in the spectral domain corresponds to an additive effects in the logarithmized spectral domain. With the assumption that H does not change with time, the effect of the channel can be normalized by computing the mean of the logarithmized spectrum and subtract it from the log-spectrum. This approach is known as mean normalization. The same reasoning also holds for the cepstral domain and the notion of *cepstral mean normalization* (CMN) is commonly used. The component-wise normalization of the variance is often paired with the mean normalization of feature vectors in ASR (Benesty et al., 2008, pp. 659). Similar to CMN, it is known as *cepstral variance normalization* (CVN) and is accomplished by first computing the variance $\boldsymbol{\sigma}_{\boldsymbol{y}}^2$ for each component of the observation vectors,

$$\boldsymbol{\sigma}_{\boldsymbol{y}}^2 = \frac{1}{T} \sum_{t=1}^{T} (\boldsymbol{y}_t - \boldsymbol{\mu}_{\boldsymbol{y}})^2, \tag{2.40}$$

and, second, by the normalization of the observation vectors,

$$\tilde{\boldsymbol{y}}_t = \frac{\boldsymbol{y}_t}{\boldsymbol{\sigma}_t}, \qquad t = 1, \ldots, T. \tag{2.41}$$

Though not motivated by a particular type of distortion, it was shown empirically that CVN increases robustness against acoustic channels, speaker variability, and additive noise (Hain et al., 1999; Benesty et al., 2008, pp. 659). While CMN and CVN address the first and second central moments, Saon et al. (2004) proposed a method that normalizes all moments for each dimension such that the transformed data is Gaussian distributed. They empirically showed that this feature-space Gaussianization can improve the performance of ASR systems in addition to other speaker-adaptive methods (see also Section 2.5).

Relative Spectral Methodology Cepstral mean normalization can be considered as a high-pass filter in time with a cutoff frequency arbitrary close to zero. This can be seen as motivation for the use of other filter approaches to increase noise robustness. With the idea that abrupt spectral changes in time are likely to be distortions with noise, Hermansky et al. (1992) proposed the *relative spectral* (RASTA) methodology, which was further developed, for example, in the work of Hermansky and Morgan (1994). It is equivalent to bandpass filtering each frequency channel through an infinite impulse response (IIR) filter with the transfer function

$$H(z) = 0.1z^4 \cdot \frac{2 + z^{-1} - z^{-3} - 2z^{-4}}{1 - pz^{-1}}, \tag{2.42}$$

where the pole p depends on the task. The pole is typically chosen in the range of 0.94 and 0.98 (Hermansky and Morgan, 1994). The high-pass filter is supposed to increase the robustness to slow variations of the channel-characteristics. The low-pass portion of the filter is supposed to suppress noise artifacts. Hermansky and Morgan (1994) also note that the choice of $p = 1$ can be seen as a special case of RASTA that resembles the mean normalization as described in the section above. In its original formulation, RASTA filtering is done by first compressing the magnitudes of the spectral values with a logarithm operation. Then each channel of the TF representation is filtered along the time axis. Afterwards, the amplitudes are expanded with an exponential function. The RASTA methodology was introduced as additional processing step during the computation of PLP features. However, as pointed out by Hermansky et al. (1992), this approach is applicable to any other method that uses a short-term spectral representation. An enhancement of the RASTA approach was presented by Morgan and Hermansky (1992) that replaces the nonparametric logarithmic nonlinearity by a one-parametric nonlinearity. The parameter of this nonlinearity was denoted as J by Morgan and Hermansky (1992) and, thus, the enhanced RASTA approach is called *J-RASTA*. With an appropriate choice of the parameter J the RASTA filtering better compensates for either convolutional or for additive noise. The drawback of J-RASTA is that optimal values for J need to be estimated for every utterance and until today there exists no standard approach for a reliable estimation.

Stereo-based Piecewise Linear Compensation for Environments While the preceding approaches CMN, CVN, and RASTA blindly transform the data, *stereo-based piecewise compensation for environments* (SPLICE) makes use of stereo data, which contains clean speech and corresponding noisy speech. It was first proposed by Deng et al. (2000) and assumes that the distribution of the noisy speech feature vectors $\boldsymbol{Y}_{1:T} \in \mathbb{R}^{d \times T}$ can be described with a *Gaussian mixture model* (GMM, Bishop (2006, pp. 430ff)),

$$p(\boldsymbol{y}) = \sum_r p(\boldsymbol{y}|r)\, p(r) \qquad \text{with} \qquad p(\boldsymbol{y}|r) = \mathcal{N}(\boldsymbol{y};\, \boldsymbol{\mu}_r, \boldsymbol{\Sigma}_r). \tag{2.43}$$

Here, $\mathcal{N}(\,\cdot\,;\, \boldsymbol{\mu}_r, \boldsymbol{\Sigma}_r)$ denotes a multivariate *Gaussian probability density function* with mean vector $\boldsymbol{\mu}$ and covariance matrix $\boldsymbol{\Sigma}$,

$$\mathcal{N}(\boldsymbol{y}; \boldsymbol{\mu}, \boldsymbol{\Sigma}) = \frac{1}{\sqrt{(2\pi)^d |\,\boldsymbol{\Sigma}\,|}} e^{-\frac{1}{2}(\boldsymbol{y}-\boldsymbol{\mu})^{\mathrm{T}}\,\boldsymbol{\Sigma}^{-1}(\boldsymbol{y}-\boldsymbol{\mu})}, \tag{2.44}$$

where d is the dimensionality of $\boldsymbol{y}$. Another assumption of SPLICE is that the *conditional probability density function* for the clean speech observation vector $\boldsymbol{x}$ given the noisy speech vector $\boldsymbol{y}$ and the region index r is Gaussian. The mean vector of the Gaussian is a linear transform of the noisy speech vector,

$$p(\boldsymbol{x}|\boldsymbol{y}, r) = \mathcal{N}(\boldsymbol{x}; \boldsymbol{A}_r\boldsymbol{y} + \boldsymbol{r}_r, \boldsymbol{\Gamma}_r). \tag{2.45}$$

For simplicity, the linear transform $\boldsymbol{A}_r$ is sometimes made to be the identity matrix. The minimum mean-squared-error estimate of the clean speech observation vector $\widehat{\boldsymbol{x}}$ is then given by (Deng et al., 2000)

$$\widehat{\boldsymbol{x}} = \boldsymbol{y} + \sum_r p(r|\boldsymbol{y})\, \boldsymbol{r}_r. \tag{2.46}$$

The correction vectors $\boldsymbol{r}_r$ and the parameters $\boldsymbol{\mu}_r$ and $\boldsymbol{\Sigma}_r$ have to be trained. Since the PDF $p(\boldsymbol{y})$ is assumed to be a mixture of Gaussians, the standard EM algorithm (Hartley, 1958) can be used to estimate $\boldsymbol{\mu}_r$ and $\boldsymbol{\Sigma}_r$. Given stereo-data in form of clean speech observation vectors $\boldsymbol{X}_{1:T} = \begin{bmatrix}\boldsymbol{x}_1 & \boldsymbol{x}_2 & \dots & \boldsymbol{x}_T\end{bmatrix} \in \mathbb{R}^{d \times T}$ and noisy speech observation vectors $(\boldsymbol{y}_1, \boldsymbol{y}_2, \dots, \boldsymbol{y}_T)$ the correction vector $\boldsymbol{r}_r$ can be estimated with a maximum likelihood criterion,

$$\boldsymbol{r}_r = \frac{\sum_t p(r|\boldsymbol{y}_n)(\boldsymbol{x}_n - \boldsymbol{y}_n)}{\sum_n p(r|\boldsymbol{y}_n)}, \tag{2.47}$$

where $p(r|\boldsymbol{y}_n)$ can be reformulated with the Bayes theorem as

$$p(r|\boldsymbol{y}_n) = \frac{p(\boldsymbol{y}_n|r)\,p(r)}{\sum_r p(\boldsymbol{y}_n|r)\,p(r)}. \tag{2.48}$$

A problem with the basic SPLICE approach is that is does not generalize well with unseen noise conditions or channel types. As presented by Droppo et al. (2001) an enhancement for SPLICE that addresses this problem is described and evaluated on the Aurora-2 task (see Appendix B.2 for the task description). The enhancement is based on an adaptive noise estimate to decrease the dependency of the SPLICE mapping on the noise statistics. While SPLICE was used to correct only the static features when it was proposed, Droppo et al. (2002) present also results where SPLICE was used to correct feature vectors together with their delta and delta-delta features.

Power-Normalized Cepstral Coefficients Kim and Stern (2009) presented a noise robust feature extraction method that they refer to as *power-normalized cepstral coefficients* (PNCC). For the time-frequency analysis a 40-channel gammatone filter bank is used. To increase noise robustness a spectral subtraction algorithm is applied that is based on the estimation of the medium-duration power of each channel. The output is then passed through a power-function nonlinearity that compresses the dynamic range of the spectral values $s_n(k)$ according to $\widehat{s}_n(k) = s_n(k)^{0.1}$. The exponent of 0.1 is replaced by 0.15 in a later work of the authors (Kim and Stern, 2010a). The estimation of the medium-duration power bias to be removed is based on the logarithmized ratio between the arithmetic to geometric mean (AM-to-GM ratio) of the running-average power $Q_n(k)$ of each channel k and each frame n. With M defining the window size, the running-average power $Q_n(k)$ is defined as

$$Q_n(k) = \frac{1}{2M+1} \sum_{n'=n-M}^{n+M} s_{n'}(k)\,. \tag{2.49}$$

The log-AM-to-GM ratio $\Psi(Q_n(k))$ for each channel k is then defined as

$$\Psi(Q_n(k)) = \log\left(\frac{1}{N}\sum_{n=1}^{N} Q_n(k)\right) - \frac{1}{N}\sum_{n=1}^{N} (\log Q_n(k))\,, \tag{2.50}$$

where N denotes the total number of frames. While the originally proposed version of the PNCC approach needs a training database for the estimation of the bias, an enhanced version of the PNCC approach was presented by Kim and Stern (2010a) that does not need a training database anymore. Besides the normalization of the filter bank output according to the 95th percentile of the short-time spectrum of the signal the key idea of the enhanced version is to choose the bias such that the AM-to-GM ratio of each channel is maximized. This maximization is referred to as "maximizing the sharpness of the power distribution" by Kim and Stern (2010a). A more detailed description about the PNCC features can be found in the thesis of Kim (2010). In this work the Matlab implementation that is publicly available from Kim and Stern (2010b) is used. After applying the power-function nonlinearity, a DCT and subsequent mean subtraction are used as final feature extraction steps for the computation of PNCCs.

2.3. Feature Transforms

Feature transforms or feature projections are commonly used in ASR systems in order to decorrelate and/or reduce the dimensionality of the feature vectors $\boldsymbol{y}$. Gales and Young (2008) give a compact overview of the feature transforms especially used in the context of automatic speech recognition. In case of a linear transform the general application is given by

$$\boldsymbol{y} = \boldsymbol{A}\tilde{\boldsymbol{y}}, \tag{2.51}$$

where $\boldsymbol{A} \in \mathbb{R}^{p \times d}$, d is the dimension of the source feature vector $\tilde{\boldsymbol{y}}$, and p is the target dimension after applying the linear transform ($p < d$). Temporal context can be included by concatenating a number of feature vectors to a composite vector and transforming the composite vector with the linear transform. If a supervised transform is to be estimated one has to decide which class labels are to be used. Labels referring to phonemes, states, or Gaussian components are typical choices here. Also, the target dimensionality has to be chosen. Haeb-Umbach and Ney (1992) did a study about the effects of different choices for the class labels. Their observation was that class labels based on phoneme states or Gaussian components lead to higher accuracies than higher level labels like phoneme labels, and that the performance when using state or component labels is similar.

In the following feature transformation methods that have found application in this work are presented. Before the transformations are described, some general notions are given: As mentioned in Section 2.1, a sequence of feature vectors $\boldsymbol{Y}_{1:T}$,

$$\boldsymbol{Y}_{1:T} = \begin{bmatrix} \boldsymbol{y}_1 & \boldsymbol{y}_2 & \dots & \boldsymbol{y}_T \end{bmatrix}, \quad \boldsymbol{Y}_{1:T} \in \mathbb{R}^{d \times T}, \tag{2.52}$$

with length T is computed for every input utterance. In the following, R denotes the total number of available utterances. The feature vector sequence that corresponds to utterance r is denoted as $\boldsymbol{Y}^{(r)}_{1:T}$,

$$\boldsymbol{Y}^{(r)}_{1:T} = \begin{bmatrix} \boldsymbol{y}^{(r)}_1 & \boldsymbol{y}^{(r)}_2 & \dots & \boldsymbol{y}^{(r)}_{T^{(r)}} \end{bmatrix}, \quad 1 \leq r \leq R, \quad \boldsymbol{Y}^{(r)}_{1:T} \in \mathbb{R}^{d \times T^{(r)}}, \tag{2.53}$$

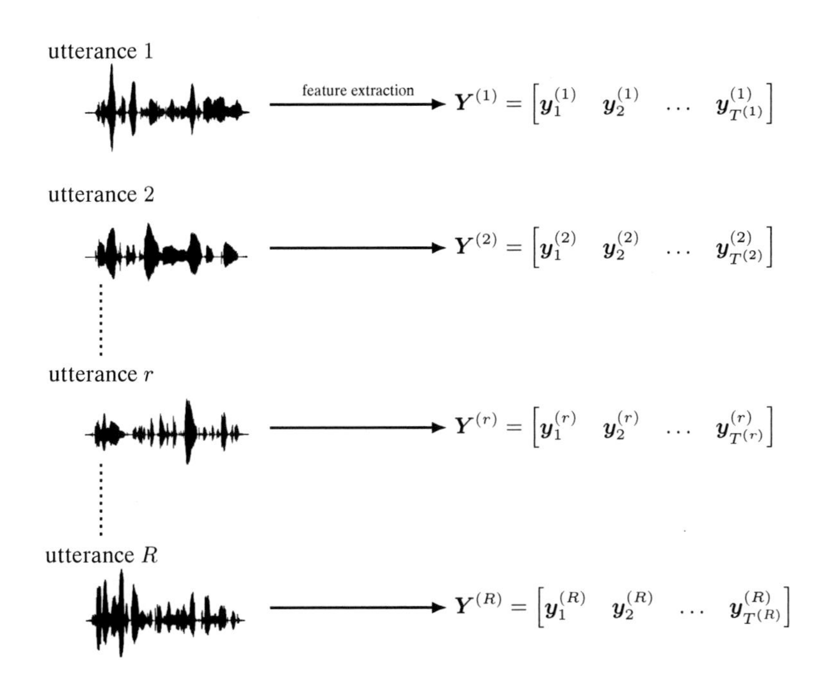

Figure 2.9.: Overview of the notions for the feature vectors and feature sequences for different utterances.

in the following. Here, the number of feature vectors for the r-th sequence is denoted as $T^{(r)}$. Figure 2.9 gives an overview of these notions. In the following, N is defined as the number of all available speech observation vectors,

$$N := \sum_{r=1}^{R} T^{(r)}, \tag{2.54}$$

and we define $\widehat{\boldsymbol{Y}} \in \mathbb{R}^{d \times N}$,

$$\widehat{\boldsymbol{Y}} := \left[\boldsymbol{Y}^{(1)} \quad \boldsymbol{Y}^{(2)} \quad \dots \quad \boldsymbol{Y}^{(r)} \quad \dots \quad \boldsymbol{Y}^{(R)}\right] \tag{2.55}$$

$$= \left[\widehat{\boldsymbol{y}}_1 \quad \widehat{\boldsymbol{y}}_2 \quad \dots \quad \widehat{\boldsymbol{y}}_N\right], \tag{2.56}$$

as the concatenation of all observation sequences to one matrix. The *global mean* $\boldsymbol{\mu}_G \in \mathbb{R}^d$ of the data set is given by

$$\boldsymbol{\mu}_G := \frac{1}{N}\sum_{n=1}^{N} \widehat{\boldsymbol{y}}_n, \tag{2.57}$$

and the *global covariance matrix* $\boldsymbol{\Sigma}_G \in \mathbb{R}^{d\times d}$ of that data set is given by

$$\boldsymbol{\Sigma}_G := \frac{1}{N}\sum_{n=1}^{N} \left(\widehat{\boldsymbol{y}}_n - \boldsymbol{\mu}_G\right)\left(\widehat{\boldsymbol{y}}_n - \boldsymbol{\mu}_G\right)^{\mathrm{T}}. \tag{2.58}$$

When considering the class labels, we define the set of all indices of the feature vectors that belong to class i as

$$\mathcal{C}_i := \{n \mid \boldsymbol{y}_n \text{ belongs to class } i,\ 1 \leq n \leq N\}, \qquad 1 \leq i \leq L, \tag{2.59}$$

where the total number of classes is denoted as L. The number of elements in $\mathcal{C}_i$ is denoted as N_i in the following. The mean $\boldsymbol{\mu}_i \in \mathbb{R}^d$ of class i is given by

$$\boldsymbol{\mu}_i := \frac{1}{N_i}\sum_{n\in\mathcal{C}_i} \boldsymbol{y}_n \tag{2.60}$$

and the covariance matrix $\boldsymbol{\Sigma}_i \in \mathbb{R}^{d\times d}$ of class i is defined as

$$\boldsymbol{\Sigma}_i := \frac{1}{N_i}\sum_{n\in\mathcal{C}_i} \left(\boldsymbol{y}_n - \boldsymbol{\mu}_i\right)\left(\boldsymbol{y}_n - \boldsymbol{\mu}_i\right)^{\mathrm{T}}. \tag{2.61}$$

2.3.1. Principal Component Analysis

Principal component analysis (PCA) is an unsupervised method for dimensionality reduction (Bishop, 2006). It is also known as "Karhunen-Loève transform". PCA estimates an orthogonal projection $\boldsymbol{A} \in \mathbb{R}^{p\times d}$, $p \leq d$, that maximizes the variance $\boldsymbol{\Sigma}_{\boldsymbol{y}}$ of the projected data $\boldsymbol{y} = \boldsymbol{A}\widetilde{\boldsymbol{y}}$. For the multi-dimensional case it can be formulated as maximizing the objective function $\mathcal{F}_{\mathrm{PCA}}$ with

$$\mathcal{F}_{\mathrm{PCA}}\left(\boldsymbol{A}\right) := |\boldsymbol{A}\boldsymbol{\Sigma}_G\boldsymbol{A}^{\mathrm{T}}|. \tag{2.62}$$

It can be shown (for example, Bishop, 2006) that Equation (2.62) is maximized by choosing $\boldsymbol{A}$ as the composite of the p eigenvectors of $\boldsymbol{\Sigma}_G$ (see Equation (2.58)) that correspond to the p largest eigenvalues.

2.3.2. Linear Discriminant Analysis

As described in the previous section, PCA does not consider class separability for determining $\boldsymbol{A}$. *Linear discriminant analysis* (LDA) as a supervised method, in contrast to PCA, does not seek a transform $\boldsymbol{A}$ that is efficient for representation, but rather for discrimination (Duda et al., 2001). The idea of LDA is to find a projection that gives a large separation between the projected class means $\boldsymbol{\mu}_i$ while also giving small within-class variances (Bishop, 2006). A *within-class scatter matrix* $\boldsymbol{\Sigma}_W$ shows the scatter of observations around their respective class means,

$$\boldsymbol{\Sigma}_W := \sum_{i=1}^{L} \frac{1}{N_i} \boldsymbol{\Sigma}_i. \tag{2.63}$$

A *between-class scatter matrix* $\boldsymbol{\Sigma}_B$ describes the scatter of the class means around the global mean,

$$\boldsymbol{\Sigma}_B := \sum_{i=1}^{L} \frac{1}{N_i} \left(\boldsymbol{\mu}_i - \boldsymbol{\mu}_G\right) \left(\boldsymbol{\mu}_i - \boldsymbol{\mu}_G\right)^{\mathrm{T}}. \tag{2.64}$$

Scalar measures of scatter are, for example, the determinants of the scatter matrices in the Equations (2.63) and (2.64). This leads to the objective function that is maximized by the LDA criterion:

$$\mathcal{F}_{\mathrm{LDA}}\left(\boldsymbol{A}\right) := \frac{|\boldsymbol{A}\boldsymbol{\Sigma}_B\boldsymbol{A}^{\mathrm{T}}|}{|\boldsymbol{A}\boldsymbol{\Sigma}_W\boldsymbol{A}^{\mathrm{T}}|}. \tag{2.65}$$

It can be shown (Fukunaga, 2001) that an optimal transformation $\boldsymbol{A}$ according to the LDA criterion is given by the eigenvectors of $\boldsymbol{S}_W^{-1}\boldsymbol{S}_B$ that correspond to the largest p eigenvalues. A common class separability criterion J_{LDA} is given by

$$J_{\mathrm{LDA}} = \mathrm{tr}(\boldsymbol{S}_W^{-1}\boldsymbol{S}_B), \tag{2.66}$$

where

$$\mathrm{tr}(\boldsymbol{X}) = x_{11} + x_{22} + \cdots + x_{nn} = \sum_{i=1}^{n} x_{ii}, \qquad \boldsymbol{X} \in \mathbb{R}^{n \times n}, \tag{2.67}$$

denotes the *trace* of the matrix $\boldsymbol{X}$. Since $\boldsymbol{A}$ is orthogonal, it is pointed out by Gales and Young (2008) that the average within-class covariance is diagonalized by an LDA, which should improve the diagonal covariance assumption often made for the acoustic models of an ASR system. Nevertheless, it is common practice, that a maximum likelihood linear transform (see Section 2.3.3) is applied subsequently to an LDA projection. The benefits of this procedure have been shown, for example, in the work of Saon et al. (2000).

There exist also other linear transforms that are often used in the field of ASR. Examples are the heterocedastic discriminant analysis (HDA) (Saon et al., 2000) or the heteroscedastic LDA (HLDA) (Kumar, 1997).

2.3.3. Maximum Likelihood Linear Transform

The *maximum likelihood linear transform* (MLLT) (Gopinath, 1998) aims at minimizing the loss in likelihood between full and diagonal covariance modeling. This transform is useful when diagonal covariance modeling constraints are given with the acoustic models. The objective function $\mathcal{F}_{\mathrm{MLLT}}$ to be maximized in the log-likelihood domain can be stated as

$$\mathcal{F}_{\mathrm{MLLT}}(\boldsymbol{A}) := \frac{1}{N}\sum_{i=1}^{L} N_i \left\{ \log \left| \mathrm{diag}\left(\boldsymbol{A}\boldsymbol{\Sigma}_i\boldsymbol{A}^{\mathrm{T}}\right) \right| - \log \left| \boldsymbol{A}\boldsymbol{\Sigma}_i\boldsymbol{A}^{\mathrm{T}} \right| \right\}. \tag{2.68}$$

The derivative of Equation (2.68) is given by

$$\begin{aligned} \frac{\partial \mathcal{F}_{\mathrm{MLLT}}(\boldsymbol{A})}{\partial \boldsymbol{A}} = & \frac{1}{N}\sum_{i=1}^{L} N_i \left\{ \boldsymbol{\Sigma}_i\boldsymbol{A}^{\mathrm{T}} \left(\mathrm{diag}\left(\boldsymbol{A}\boldsymbol{\Sigma}_i\boldsymbol{A}^{\mathrm{T}}\right)\right)^{-1} \right. \\ & \left. - \boldsymbol{\Sigma}_i\boldsymbol{A}^{\mathrm{T}} \left(\boldsymbol{A}\boldsymbol{\Sigma}_i\boldsymbol{A}_{[p]}^{\mathrm{T}}\right)^{-1} \right\}. \end{aligned} \tag{2.69}$$

A solution for the MLLT criterion can be computed with numerical optimization methods.

2.4. Hidden Markov Model Speech Recognition Systems

This section introduces the notions and methods that are needed for the design of a speech recognition system based on hidden Markov models. The modeling of the acoustic realizations is described, as well as the incorporation of grammatical knowledge about the used language. Furthermore, solutions for general problems of this modeling approach and the decoding of a speech signal are described in this section.

2.4.1. Acoustic Models

As already mentioned in the introduction, a *phoneme* can be though of as an ideal sound unit with a complete set of articulatory gestures (Deller et al., 1993). Each phoneme represents a class of sounds that have the same meaning. In contrast, a *phone* denotes the actual sound that is produced in speaking. Each phoneme has a set of associated phones, so-called *allophones*, that are thought of as being categorically the same. There are standardized phonemic and phonetic transcription alphabets. The *international phonetic alphabet* (IPA) (Ladefoged, 2001) is the world-wide standard for phonetic notation in general. The *Arpabet* (Deller et al., 1993) maps the characters defined in the IPA to ASCII[1] characters. As an example, the word

"dictionary"

can be transcribed with the Arpabet as

"/d/ /ih/ /k/ /sh/ /ah/ /n/ /eh/ /r/ /iy/".

The dictionary of an ASR system, which is often also called "vocabulary", contains a list of all recognizeable words together with their corresponding transcriptions to their acoustic events. ASR systems with a vocabulary smaller than about 1000 words are usually referred to as "small vocabulary" ASR systems. An ASR system with a vocabulary with 1000 to 10000 words is referred to a "medium-size vocabulary" system, and a vocabulary with more than 10000 words is usually referred to as "large vocabulary" system. Statistical speech recognition systems model speech as temporal sequences of individual acoustic events. Whether such an acoustical event is a whole word or an individual phoneme depends on the application of the ASR system. *Hidden Markov models* (HMM) are mostly used for acoustic modeling in the field of ASR systems nowadays. Extensions to the HMM approach or alternative frameworks for acoustic modeling are an active field of research (see, for example, Mohri et al., 2002; Henter and Kleijn, 2011; Seide et al., 2011). Since the modeling approach described in the following is still part of the vast majority of state-of-the-art ASR systems, these alternatives are not further considered in this work. In the following the principles of HMMs themselves and their application for ASR are briefly explained. Detailed descriptions can be found in the book of Bishop (2006) and, specifically for the task of ASR, in the books of Rabiner and Juang (1993) and Gales and Young (2008). A basic introduction to the use of HMMs for ASR is given in the article of Rabiner and Juang (1986).

[1] American Standard Code for Information Interchange, a set of digital codes representing letters, numerals, and other symbols, widely used as a standard format in the transfer of text between computers (Dic, 2010).

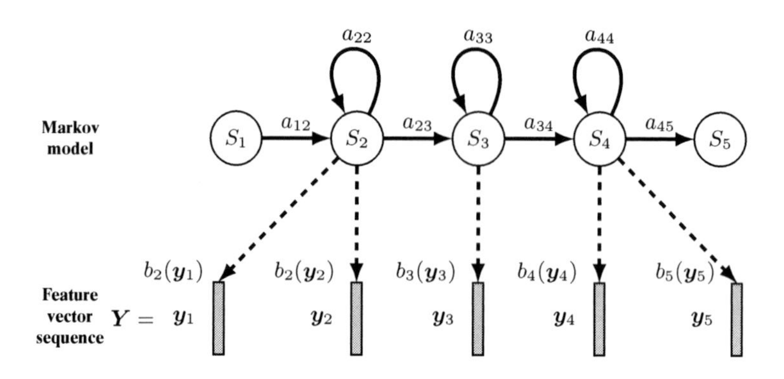

Figure 2.10.: Left-to-right HMM-based phoneme model without state skips and with an exemplary feature vector sequence $\boldsymbol{Y}_{1:5}$. Adapted from Benesty et al. (2008).

A HMM is a doubly stochastic process with an underlying stochastic process that is not observable (that means it is hidden), but can only be observed through another set of stochastic processes that produce the sequence of observations (Rabiner and Juang, 1986). The model consists of a number of elements and is schematically shown in Figure 2.10. With Q denoting a finite number of states of an HMM, the individual states are denoted as $S_1, S_2, \dots, S_Q$. States are changed from state i to state j according to given transition probability distributions and are denoted as a_{ij} in the following. The set of all transition probabilities is denoted as

$$\mathcal{A} := \{\, a_{ij} \,|\, 1 \leq i, j \leq Q \,\}. \tag{2.70}$$

After each transition, an observation output is generated according to an observation probability distribution that depends only on the current state. The observation probability distribution of state S_j is denoted as $b_j(\,\cdot\,)$ in the following and usually multivariate GMMs are used in the field of ASR,

$$b_j(\boldsymbol{y}_t) = \sum_{m=1}^{M} c^{(jm)} \mathcal{N}(\boldsymbol{y}_t; \boldsymbol{\mu}^{(jm)}, \boldsymbol{\Sigma}^{(jm)}), \quad 1 \leq j \leq Q, \tag{2.71}$$

where M is the number of mixture components in state j, $c^{(jm)}$ is the weight of the m-th component and $\mathcal{N}(\,\cdot\,; \boldsymbol{\mu}, \boldsymbol{\Sigma})$ is a multivariate Gaussian probability density function with mean vector $\boldsymbol{\mu}$ and covariance matrix $\boldsymbol{\Sigma}$ (as given in Equation (2.44)). Besides full covariance matrices, the covariance matrices can be constrained to be of diagonal or block-diagonal form in order to reduce the number of model parameters. Commonly, diagonal covariance matrices are used for ASR. The weighting coefficients $c^{(jm)}$ must be nonnegative and satisfy

$$\sum_{m=1}^{M} c^{(jm)} = 1, \quad 1 \leq j \leq Q. \tag{2.72}$$

The set of all observation probability distributions is denoted as

$$\mathcal{B} := \{ b_j \,|\, 1 \leq j \leq Q \}. \tag{2.73}$$

For ease of notation, the acoustic model parameters are summarized with

$$\boldsymbol{\lambda} := \{\mathcal{A}, \mathcal{B}\} \tag{2.74}$$

in the following. If the vocabulary to be recognized is small, each word of the vocabulary could be modeled with an individual HMM. For large vocabularies, ASR systems use pronunciation dictionaries to map phoneme sequences to whole words, where pronunciation variations may also be provided by the dictionary. With words being concatenations of phonemes, a decoder of an ASR systems concatenates the corresponding HMMs of the phonemes to form words during decoding. A typical topology for phoneme-level HMMs is a three-state, left-to-right model without state skips as illustrated in Figure 2.10. To circumvent the problem of normalization of transition probabilities in the last state, a nonemitting state is usually put at the beginning and the end of the HMM, which can also be seen in Figure 2.10.

Three main problems arise in the application of HMMs in practice (Rabiner and Juang, 1986):

1. **Probability evaluation:** Given an observation sequence $\boldsymbol{Y}_{1:T}$ and the model $\boldsymbol{\lambda}$, how do we compute $p(\boldsymbol{Y}_{1:T} \,|\, \boldsymbol{\lambda})$, the probability of the observation sequence, given the model?

2. **Optimal state sequence:** Given an observation sequence $\boldsymbol{Y}_{1:T}$ and the the model $\boldsymbol{\lambda}$, how do we choose a state sequence that best "explains" the observations?

3. **Parameter estimation:** How do we adjust the parameters $\boldsymbol{\lambda}$ to maximize the observation probability given $\boldsymbol{\lambda}$, $p(\boldsymbol{Y}_{1:T}|\boldsymbol{\lambda})$?

Common solutions to each of the three problems are described in the following:

Probability evaluation The first problem is about the computation of the probability of an observation sequence $\boldsymbol{Y}_{1:T} = (\boldsymbol{y}_1, \boldsymbol{y}_2, \ldots, \boldsymbol{y}_T)$ given a model $\boldsymbol{\lambda}$, that is, $p(\boldsymbol{Y}_{1:T} \mid \boldsymbol{\lambda})$. Let

$$\boldsymbol{\theta} = (\theta_1, \theta_2, \ldots, \theta_T), \quad \theta_t \in \{1, 2, \ldots, Q\}, \tag{2.75}$$

denote a state sequence with one corresponding state for each observation. A direct approach for the computation of $p(\boldsymbol{Y}_{1:T} \mid \boldsymbol{\lambda})$ would sum the joint probability of $\boldsymbol{Y}$ and $\boldsymbol{\theta}$ given the model $\boldsymbol{\lambda}$ over all possible state sequences with length T,

$$p(\boldsymbol{Y}_{1:T} \mid \lambda) = \sum_{\boldsymbol{\theta}} p(\boldsymbol{Y}, \boldsymbol{\theta} \mid \lambda) \tag{2.76}$$

$$= \sum_{\boldsymbol{\theta}} p(\boldsymbol{Y} \mid \boldsymbol{\theta}, \lambda) \cdot p(\boldsymbol{\theta} \mid \lambda) \tag{2.77}$$

$$= \sum_{\theta_1, \theta_1, \ldots, \theta_T} b_{\theta_1}(\boldsymbol{y}_1) a_{\theta_1, \theta_2} b_{\theta_2}(\boldsymbol{y}_2) a_{\theta_2, \theta_3} \cdots a_{\theta_{T-1}, \theta_T} b_{\theta_T}(\boldsymbol{y}_T). \tag{2.78}$$

There are Q^T possible state sequences and it can be seen that the complexity given by Equation (2.76) is of order $\mathcal{O}(2T \cdot Q^T)$. To make the computation of $p(\boldsymbol{Y}_{1:T} \mid \boldsymbol{\lambda})$ more feasible, the *forward procedure* can be used and is explained in the following: Let

$$\alpha_t(i) := p(\boldsymbol{y}_1, \boldsymbol{y}_2, \ldots, \boldsymbol{y}_t, \theta_t = i \mid \boldsymbol{\lambda}) \tag{2.79}$$

define the forward variable, which is the probability of the (partial) sequence $\boldsymbol{Y}_{1:t}$ and being in state i at time t, given the model $\boldsymbol{\lambda}$. Thus, $\alpha_t(i)$ is often also denoted as *forward probability*. Solving for $\alpha_t(i)$ inductively yields

$$\alpha_1(i) = b_i(\boldsymbol{y}_1), \quad 1 \le i \le Q, \tag{2.80}$$

and

$$\alpha_{t+1}(j) = \left[\sum_{i=1}^{Q} \alpha_t(i) a_{ij}\right] b_j(\boldsymbol{y}_{t+1}), \quad 1 \le t \le T-1, \quad 1 \le j \le Q. \tag{2.81}$$

The desired computation of $p(\boldsymbol{Y}_{1:T} \mid \boldsymbol{\lambda})$ is then given with

$$p(\boldsymbol{Y}_{1:T} \mid \boldsymbol{\lambda}) = \sum_{i=1}^{Q} \alpha_T(i). \tag{2.82}$$

Alternatively, a *backward variable* $\beta_t(i)$ can be defined with

$$\beta_t(i) := p(\boldsymbol{y}_{t+1}, \boldsymbol{y}_{t+2}, \ldots, \boldsymbol{y}_T \mid \theta_t = i, \boldsymbol{\lambda}), \tag{2.83}$$

which is the probability of the partial observation from $t+1$ to T, given state i at time t and the model $\boldsymbol{\lambda}$. Similar to the forward variable, $\beta_i(t)$ is also often denoted as the *backward probability* and is computed iteratively,

$$\beta_T(i) := 1, \quad 1 \leq i \leq Q \tag{2.84}$$

and

$$\beta_t(i) := \sum_{j=1}^{Q} a_{ij} b_j\left(\boldsymbol{y}_{t+1}\right) \beta_{t+1}(j), \quad t = T-1, T-2, \ldots, 1, \quad 1 \leq i \leq Q. \tag{2.85}$$

It can be seen that the complexity of the forward procedure and of the backward procedure is of order $\mathcal{O}(Q^2T)$, which, in practice, yields several orders of magnitude less computations in comparison to the direct approach from Equation (2.76).

Optimal state sequence The problem of finding the optimal state sequence for an observation sequence $\boldsymbol{Y}_{1:T}$ is to find the state sequence $(\theta_1, \theta_2, \ldots, \theta_T)$ that best explains a given observation sequence. In case of HMM compositions as described above a solution of this problem would, for example, yield the most likely word sequence for a given utterance. The single best state sequence maximizes $p\left(\theta_1, \theta_2, \ldots, \theta_T,\ \boldsymbol{Y}_{1:T} \mid \boldsymbol{\lambda}\right)$. For solving this problem efficiently, a dynamic programming approach can be used that is known as the *Viterbi algorithm* (Viterbi, 1967): Let

$$\delta_t(i) := \max_{\theta_1, \ldots, \theta_{t-1}} p\left(\theta_1, \theta_2, \ldots, \theta_{t-1},\ \theta_t = i,\ \boldsymbol{Y}_{1:t} \mid \boldsymbol{\lambda}\right) \tag{2.86}$$

describe the highest probability along a single path up to time t that ends in state i. The Viterbi algorithm iteratively computes

$$\delta_{t+1}(j) = \left[\max_i \delta_t(i) a_{ij}\right] \cdot b_j(\boldsymbol{y}_{t+1}), \tag{2.87}$$

while keeping track of the argument that maximized Equation (2.87) for each t and i. The forward procedure described above is similar in its computation. The main difference is that the Viterbi algorithm maximizes over previous states, instead of the summing operation. Thus, the computational complexity is also of order $\mathcal{O}(Q^2T)$.

Parameter estimation The acoustic model parameters $\boldsymbol{\lambda} = \{\mathcal{A},\ \mathcal{B}\}$ can be estimated with the *Baum-Welch algorithm* (Baum et al., 1970), which is a special case of the *expectation-maximization* (EM) algorithm (Hartley, 1958). For GMM densities, which are commonly used in ASR systems, the parameters to be (re-)estimated are the mean vectors $\boldsymbol{\mu}^{(jm)}$ and covariance matrices $\boldsymbol{\Sigma}^{(jm)}$, as well as the weighting coefficients $c^{(jm)}$ of the M Gaussian densities for each state j of the HMMs (compare with Equation (2.71)) and the transition probabilities $\mathcal{A}$.

Let $\gamma_t^{(k)}(j)$ be the probability of being in state j at time t with the k-th mixture component accounting for $\boldsymbol{y}_t$ (Rabiner and Juang, 1993),

$$\gamma_t^{(k)}(j) = \left[\frac{\alpha_t(j)\beta_t(j)}{\sum_{j=1}^{Q} \alpha_t(j)\beta_t(j)}\right]\left[\frac{c^{(jk)}\mathcal{N}(\boldsymbol{y}_t; \boldsymbol{\mu}^{(jk)}, \boldsymbol{\Sigma}^{(jk)})}{\sum_{m=1}^{M} c^{(jm)}\mathcal{N}(\boldsymbol{y}_t; \boldsymbol{\mu}^{(jm)}, \boldsymbol{\Sigma}^{(jm)})}\right]. \tag{2.88}$$

The use of the forward and backward variables to estimate a frame/state alignment $\gamma_t^{(k)}(j)$ is also known as the *forward-backward algorithm*. Then, the reestimation formulas for the weighting coefficients, mean vectors, and for the covariance matrices are of the form

$$\widehat{c}^{(jk)} = \frac{\sum_{t=1}^{T} \gamma_t^{(k)}(j)}{\sum_{t=1}^{T}\sum_{k=1}^{M} \gamma_t^{(k)}(j)}, \tag{2.89}$$

$$\widehat{\boldsymbol{\mu}}^{(jk)} = \frac{\sum_{t=1}^{T} \gamma_t^{(k)}(j) \cdot \boldsymbol{y}_t}{\sum_{t=1}^{T} \gamma_t^{(k)}(j)}, \tag{2.90}$$

$$\widehat{\boldsymbol{\Sigma}}^{(jk)} = \frac{\sum_{t=1}^{T} \gamma_t^{(k)}(j) \cdot (\boldsymbol{y}_t - \boldsymbol{\mu}^{(jk)})^{\mathrm{T}}(\boldsymbol{y}_t - \boldsymbol{\mu}^{(jk)})}{\sum_{t=1}^{T} \gamma_t^{(k)}(j)}. \tag{2.91}$$

In order to reestimate the transition probabilities, we define $\xi_t(i,j)$ as the probability of being in state i at time t and being in state j at time $t+1$,

$$\begin{aligned}\xi_t(i,j) &= P(\theta_t = i,\, \theta_{t+1} = j \,|\, \boldsymbol{Y},\, \boldsymbol{\lambda}) \\ &= \frac{p(\theta_t = i,\, \theta_{t+1} = j,\, \boldsymbol{Y},\, |\, \boldsymbol{\lambda})}{p(\boldsymbol{Y} \,|\, \boldsymbol{\lambda})}.\end{aligned} \tag{2.92}$$

With the forward and backward variables from Equations (2.81) and (2.85), respectively, $\xi_t(i,j)$ can be written as

$$\xi_t(i,j) = \frac{\alpha_t(i)a_{ij}b_j(\boldsymbol{y}_{t+1})\beta_{t+1}(j)}{\sum_{k=1}^{Q}\sum_{l=1}^{Q} \alpha_t(k)a_{kl}b_l(\boldsymbol{y}_{t+1})\beta_{t+1}(l)}. \tag{2.93}$$

Now, the transition probabilities can be reestimated as the ratio between the expected number of transitions from state i to state j and the expected number of transitions from state i,

$$\hat{a}_{ij} = \frac{\sum_{t=1}^{T-1} \xi_t(i,j)}{\sum_{t=1}^{T-1}\sum_{k=1}^{Q} \xi_t(i,k)}. \tag{2.94}$$

With a given initialization, the parameters $\boldsymbol{\lambda}$ can be iteratively estimated with the equations shown above. A common approach for the initialization of the parameters $\boldsymbol{\lambda}^{(0)}$ is known as the *flat start procedure*, which assigns the global mean and global covariance of the feature vectors to each component of the output distributions of the HMMs and sets all transition probabilities to be equal.

Monophone and Triphone Modeling Acoustic models that represent a single phoneme without considering the preceding and following phoneme are referred to as monophone models or *monophones*. Since speech is the product of continuously moving articulators (see Section 1.2), a monophone modeling does not take into account that coarticulation leads to a considerable large amount of variability for a single phoneme in different phonetic contexts. A common approach to capture coarticulatory effects is to use an individual model for every possible combination of left and right neighboring phonemes. The resulting context-dependent acoustic models are referred to as triphone models or *triphones*. An example for the relation between monophone and triphone modeling for the word "great" (according to CMU (2011)) would be

word:	g r e a t,
monophone:	/g/ /r/ /ey/ /t/,
triphone:	/sil-g+r/ /g-r+ey/ /r-ey+t/ /ey-t+sil/.

Here, the notation from the *Hidden Markov Model Toolkit* (HTK) (Young et al., 2009) is used, where $/x - q + y/$ denotes the triphone with base phoneme q, the preceding phoneme x, and y as its following phoneme. "sil" denotes a hypothetical silence in the above example. With N base phonemes, there are theoretically N^3 triphones. To overcome the potential problem of data sparsity the number of parameters of the acoustic models can be reduced by sharing parameters; Gales and Young (2008) use the notion of *logical* triphones, which can be mapped to a reduced set of *physical* models by clustering and tying together the parameters. There exist different methods to accomplish this: *Data-driven clustering* (for example, Lee et al., 1989; Young et al., 2009) merges the parameters of models with a certain distance measure that is dependent on the type of the used state distribution. A disadvantage of this approach is that it does not handle the problem of unseen triphones in the training data. *Decision-tree based clustering* makes use of a predefined phonetic question set and a decision tree to group triphones with similar properties. With the rules being independent from the available training data, this approach defines a model for every possible triphone. Starting from the root of the tree, yes-no questions assign a unique leaf-node to each triphone state. Exemplary questions are of the form "Is phoneme to the left a nasal?" or "Is the phoneme to the right a glide?" (Young et al., 2009). The states within the same leaf-node then share the same parameters. Using single Gaussian distributions, an optimal decision tree is constructed with the CART approach (Breiman et al., 1984) in a maximum-likelihood sense. The question set to choose from has to be defined before the construction of the decision tree. Bahl et al. (1991); Young et al. (1994) discuss the problem of defining suitable question sets. For the construction of a decision tree, a question is iteratively selected that gives the largest increase in likelihood for each node.

2.4.2. Generic Training Procedure for Acoustic Model Parameters

The estimation of the parameters for the acoustic models has been formally described in Section 2.4.1 and is accomplished with the EM algorithm. The results of the parameter estimations depend on their initialization. To mitigate the problem of local maxima during the training process, commonly an iterative procedure is followed that starts with more general models and refines these in every iteration. An exemplary procedure is described, for example, by Young et al. (2009, Chapter 3). A brief overview of this procedure is given in the following:

1. In a first step a monophone set is created with output probability density functions given by single Gaussians. Commonly, a three state left-to-right HMM topology is used as depicted in Figure 2.10 on page 40. If corresponding segmental information is available the mean and covariance parameters can be individually initialized. Alternatively, a flat-start initialization can be used, in which means and covariances are equal to the global mean and covariance of the training data. The initial transition probabilities in the recipe of Young et al. (2009) are set to 0.6 and 0.4 for staying in the current state, and for switching to the next state, respectively. Given this initialization, the parameters are re-estimated with three to four iterations of EM.

2. In the second step triphone models are initialized on the basis of the monophone models from the first step. All possible triphones $/x - q + y/$ are initialized by cloning the parameters of the monophone models corresponding to $/q/$. After a re-estimation of the parameters with EM, a decision tree as described in Section 2.4.1 is generated for each state in each model. These trees are used to cluster the state parameters of the triphones, which leads to so-called "tied-state triphone models". After the tying EM is used again for a re-estimation of the parameters.

3. The third step involves iteratively increasing the number of Gaussian mixture components by finding the mixture with the largest weight and splitting it. The optimal number of mixtures depends on the task, the modeling constraints,

and on the amount of available training data. If a development set is available, the number of mixture components can be determined by observing the recognition performance on the development data.

2.4.3. Language Models

Linguistic knowledge is represented within ASR systems with so-called *language models*. They provide an estimate for the probability $P(\boldsymbol{W})$ (compare with Equation (2.3)), which is the probability for the occurrence of a word sequence $\boldsymbol{W}$ of length R,

$$\boldsymbol{W}_{1:R} = (w_1, w_2, \ldots, w_R). \tag{2.95}$$

For ease of notation the subscript of $\boldsymbol{W}_{1:R}$ is omitted if appropriate in the following. The prior probability $P(\boldsymbol{W})$ is given by

$$\begin{aligned} P(\boldsymbol{W}) &= P(w_1, w_2, \ldots, w_R) \\ &= P(w_R \mid w_1, w_2, \ldots, w_{R-1}) \cdot P(w_1, w_2, \ldots, w_{R-1}) \\ &= \prod_{k=1}^{R} P(w_k \mid w_1, \ldots, w_{k-1}). \end{aligned} \tag{2.96}$$

To circumvent the need for a reliable estimate of the conditional probabilities for all word sequences of a language, ASR systems usually truncate the word history in Equation (2.96) to a certain number of preceding words. The generated language model

$$P(\boldsymbol{W}) = \prod_{k=1}^{R} P(w_k \mid w_{k-N+1}, w_{k-N+2}, \ldots, w_{k-1}) \tag{2.97}$$

considers $N-1$ preceding words and is called an *N-gram* language model. For the special cases $N = 2$ and $N = 3$ the terms “bigram language model” and “trigram language model”, respectively, are commonly used. When no history of preceding words is taken into account (in case of $N = 1$) the language model is called “unigram”. If the language model does not consider any word counts and assumes equal probabilities for all words it is called “zero-gram language model”.

Maximum-likelihood estimates of the N-gram probabilities are estimated from training texts by counting the occurrences. As an example, the probability $P(w_{k-1}, w_k)$ of a bigram language model would be estimated as following: Let $C(w_{k-1}, w_k)$

be the number of occurrences of the word sequence w_{k-1}, w_k and similarly for $C(w_{k-1})$. Then $P(w_{k-1}, w_k)$ can be estimated as

$$P(w_k \mid w_{k-1}) \approx \frac{C(w_{k-1}, w_k)}{C(w_{k-1})}. \tag{2.98}$$

The major difficulty in language model estimation is the problem of data sparsity. To allow for the recognition of word sequences that have not been observed in the training texts it is necessary to reserve some probability mass for these. This procedure is also known as "smoothing". A variety of methods has been proposed and an overview of these techniques together with a comparison is given by Chen and Goodman (1999). Standard methods involve *discounting schemes*, first proposed by Katz (1987), which redistribute the probability mass of observed word sequences to unseen ones. In addition, *back off schemes* make use of less precise, but more frequently observed lower-order N-grams.

Alternatively, class-based language models can be used in which every word w_k is mapped to a corresponding word-class ω_k^W (Brown et al., 1992). Similar to Equation (2.97), it is then

$$P(\boldsymbol{W}) = \prod_{k=1}^{R} P(w_k \,|\, \omega_k^W) \cdot p(\omega_k^W \,|\, \omega_{k-N+1}^W, \omega_{k-N+2}^W, \ldots, \omega_{k-1}^W). \tag{2.99}$$

Since the number of word classes is much smaller than the number of word-based N-grams, the class N-gram probabilities can be estimated more reliable. The word-classes can be chosen iteratively with a maximum-likelihood criterion (Martin et al., 1995).

2.4.4. Decoding, Scoring, and Significance

Decoding Given a sequence of observations $\boldsymbol{Y}$, the *decoder* of an ASR system is supposed to return the most likely word sequence $\widehat{\boldsymbol{W}}$. As described in detail in Section 2.4.1, the Viterbi algorithm can be used to efficiently compute the state sequence that was most likely to have generated the observations $\boldsymbol{Y}$. Without any further constraints, a direct implementation of the Viterbi algorithm for an ASR task becomes unfeasible in practice. Two reasons for this are N-gram language models and cross-word triphone contexts, both of which greatly expand the search space (Gales and Young, 2008, Chapter 2.4). Various approaches have been proposed to solve this problem and an overview of these can be found in the book of Gales and Young (2008, Chapter 2.4). The method that is described next makes the HMM topologies explicit by constructing a recognition network. For a general large-vocabulary recognition system, the network represents all words in parallel in a loop. The words themselves are represented as sequences of phoneme models given by the dictionary as described above.

A hypothesis consists of a path through the network and has an assigned log-likelihood. An explicit representation of such a path can be made by introducing the notion of a *token* (Young et al., 1989), which consists of a log-likelihood value and a pointer to history information. The token-passing algorithm passes a start token from node to node and updates the token likelihood at each transition, while tracking the individual likelihoods and history information. If a token is propagated between two words, often the probability given by the language model is scaled by a *grammar scale factor* α and a word-insertion penalty β is added. Thus, in the logarithmic domain the total likelihood is computed as

$$\log p(\boldsymbol{W} \mid \boldsymbol{Y}) = \log p(\boldsymbol{Y} \mid \boldsymbol{W}) + \alpha P(\boldsymbol{W}) + \beta|\boldsymbol{W}|. \quad (2.100)$$

In the literature the weighted log-likelihood in Equation (2.100) is often also referred to as *score* or *confidence* (see, for example, Young et al., 2009). At the end of the utterance the word sequence is chosen that corresponds to the token with the highest log-likelihood.

Efficiency with this approach is achieved by deleting tokens whose probabilities fall below a certain threshold. Therefore, the threshold, which is also called *beam-width* in this context, is the difference in log-likelihood to the most likely path in the network. Every token with a likelihood below this threshold is removed during the decoding process. This thresholding approach is also known as *beam search*. A smaller beam-width decreases the search space of possible word sequences and increases the speed of the decoding system. However, with a too small beam-width the most likely path might be pruned during the decoding, which would yield a search error. Thus, deciding for an optimal beam-width is a trade-off between efficiency and error rate in practice.

Instead of choosing only the token with the highest likelihood at the end of the Viterbi algorithm, a certain number of most likely paths could be stored as result of a decoding process. Usually, a *lattice* representation of these paths is used. A lattice consists of nodes and arcs. The nodes represent points in time, while the arcs represent word hypotheses. Lattices are generated with the token-passing algorithm and can be used for different purposes. One application is *lattice rescoring*. With lattices being constrained recognition networks, a Viterbi decoder can be used to rescore the acoustic and language model factors assigned to the nodes of the lattice. This allows a more rapid evaluation of alternative, similar ASR systems.

Scoring One way for quantifying the performance of an ASR system is given in terms of the *word error rate* (WER). Once the decoder has computed a transcription for a given utterance, a reference (ground truth) transcription is compared with the output of the decoder. To determine, which segments are compared with each other, an optimal string match is computed with a dynamic programming approach. The string matching minimizes the Levenshtein distance (Levenshtein, 1966) with different penalties for insertions, deletions, and substitutions[2]. Given optimal alignments for all test utterances, the WER in units of percentage is computed as

$$\text{WER} = 100 \times \left(1 - \frac{N - D - S - I}{N}\right) \quad [\%], \tag{2.101}$$

where N is the total number of words, D the number of deletions, S the number of substitution, and I the number of insertions, respectively. Complementary, the *word accuracy* (WA) is computed as

$$\text{WA} = 100 \times \left(\frac{N - D - S - I}{N}\right) = 100 - \text{WER} \quad [\%]. \tag{2.102}$$

In case of phoneme recognition one speaks of "phoneme error rate" and "phoneme accuracy", respectively.

Significance In order to compare the performances of two different ASR systems with each other, a measure of significance of the relative improvement is useful. A significance test is a statistical test for the null hypothesis that a system A is identical to a system B. In this work, the *matched pairs sentence-segment word error* (MPSSWE) test as introduced by Gillick and Cox (1989) is used. The following description is based on the one given by Nat (2011).

The overall aim is to determine whether two given ASR systems perform significantly different or not. The MPSSWE test splits the corresponding transcriptions of both systems into error segments. The error segments are bounded on both sides by words correctly recognized by both systems or by the beginning or end of the utterances. As an example, Table 2.1 shows the outputs of systems A and B together with the reference transcription and their alignment. Overall, three segments with errors can be seen. In the first segment, system A deletes "is" and substitutes "it" by "its", while system B does not make any errors. In the second segment, system A deletes "a" and, again, system B does not make an error. In segment three, system B substitutes "day" by "way" and system A makes no error. The segmentation is computed with dynamic programming and approximately validates the independence assumption that is needed for the significance test. With a given alignment the number of errors N_A^i and N_b^i for each of the systems A and B and for each segment i can

[2]In this work, the implementation contained in HTK (Young et al., 2009) was used, which uses the penalty weights 7, 7, and 10 for insertions, deletions, and substitutions, respectively.

Table 2.1.: Exemplary alignment and error segments of matched pairs sentence-segment word error (MPSSWE) test

segment:	1	2		3	
reference:	it is	a	great	day	for shopping
system A:	ITS		great	day	for shopping
system B:	it is	a	great	WAY	for shopping

be computed. The estimated mean of the error differences $\tilde{\mu}_Z$ between the two systems over all segments is given by

$$\tilde{\mu}_Z = \frac{1}{n} \sum_{i=1}^{n} N_A^i - N_B^i, \tag{2.103}$$

where n is the total number of error segments. With

$$\tilde{\sigma}_Z^2 = \frac{1}{n-1} \sum_{i=1}^{n} \left(N_A^i - N_B^i - \tilde{\mu}_Z \right)^2 \tag{2.104}$$

being the estimate of variance of the error differences, let $\hat{\mu}_Z$ be

$$\hat{\mu}_Z = \frac{\tilde{\mu}_Z}{(\sigma_Z / \sqrt{n})}. \tag{2.105}$$

If n is large enough, then $\hat{\mu}_Z$ will approximately have a Normal distribution with unit variance (Gillick and Cox, 1989). Here, the null hypothesis asserts that the distribution of $\tilde{\mu}_Z$ has zero mean. By rejecting the null hypothesis it is concluded that system A is significantly different from system B. The null hypothesis is rejected, if

$$\varphi := P(|Z| \geq |\hat{\mu}_Z|) \leq 0.05, \tag{2.106}$$

where Z is a random variable with distribution $\mathcal{N}(\,\cdot\,; 0, 1)$. In this work, a *significance level* of 0.05 is used, which is a typical choice for MPSSWE (for example, Sainath et al., 2011). If φ from Equation (2.106) is larger than 0.05, the two considered systems A and B are considered to be identical. If φ is equal or smaller than 0.05 the two systems are considered to produce a significant amount of different errors and, thus, are considered to be not identical. An implementation of MPSSWE from the National Institute of Standards and Technology (NIST) is available in the *Speech Recognition Scoring Toolkit* (SCTK) (Pallet et al., 1990; Nat, 2010).

2.5. Speaker Compensation Methods

Inter- and intra-speaker variabilities (see also Section 1.2) have considerable negative effects on the recognition performance. Because of that techniques that try to compensate for these variabilities are commonly used in state-of-the-art ASR systems nowadays. From a pattern recognition point of view, speaker compensation has the effect of lessening the mismatch in feature space between training and test data. If the decoding process is not limited to a single ASR system a clustering approach can be used: The training and test data could be clustered according to certain information inferred from the utterances. For example, given the gender of the training speakers, gender-dependent acoustic models could be trained. Before the recognition, the genders of the speakers need to be determined. The model set that corresponds to the determined gender is then used for recognition. Besides gender-dependent clustering, more general clustering schemes have been proposed. Gao et al. (1997), for example, define clusters according to a distance measure that depends on the parameters of acoustic models. While the clustering approach makes use of several acoustic model sets, the normalization and adaptation methods that are explained in the following are applied within individual ASR systems. An exhaustive list of all normalization and adaptation schemes used in practice nowadays is beyond the scope of this work. However, key concepts of commonly used methods that where also used within the experiments of this work are described here.

With respect to their location within the processing chain of an ASR system (see also Figure 2.1 on page 14), speaker adaptation and normalization methods can be grouped into three classes of methods. The first class performs a normalization of the features during or after they have been extracted. These methods are usually referred to as *normalization* methods. The second class of methods adapts the parameters of the acoustic models to the characteristics of the individual speakers and are usually referred to as *adaptation* methods. Finally, the third group of methods extracts features with transforms that are (ideally) invariant to the variabilities introduced by the speakers. Common normalization and adaptation methods are described in the following. Methods for the extraction of invariant features are presented in detail in Chapter 4.

2.5.1. Vocal Tract Length Normalization

One inter-speaker variability (see Section 1.2) with a major impact on the performance of ASR systems is the vocal tract length. In a simple form, the vocal tract shape can be considered as a lossless, uniform acoustic tube with a certain

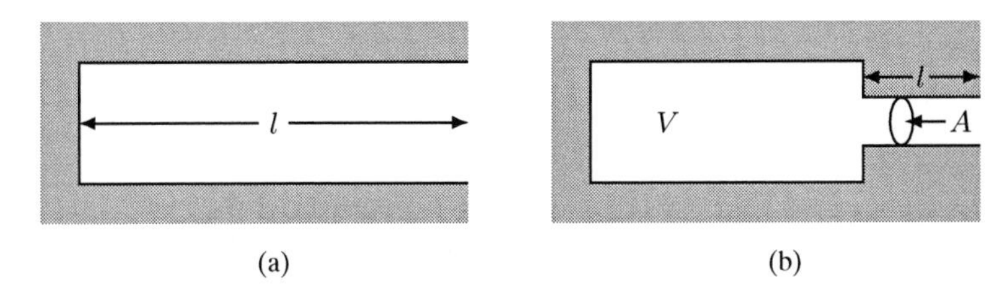

Figure 2.11.: (a) Uniform tube model with length l, (b) Helmholtz resonator with volume V of the back cavity, A being the cross-sectional area, and the length l of the narrow tube (Eide and Gish, 1996).

length l (Deller et al., 1993, Chapter 3) as illustrated in Figure 2.11 (a). With one end closed, and the other end opened, the resonance frequencies (the formants) of this acoustic tube occur at frequencies F_i,

$$F_i = \frac{c}{4l}(2i-1), \quad i = 1, 2, 3, \ldots, \tag{2.107}$$

where $c = 350$ m/sec is the speed of sound in moist air at a temperature of 37°C (Deller et al., 1993). From Equation (2.107) can be seen that a change in vocal tract length leads to a linear scaling of the formant frequencies. While the uniform tube model is appropriate for relative open vowels such as /AA/ (Eide and Gish, 1996), a Helmholtz resonator as illustrated in Figure 2.11 (b) is a better approximation of the vocal tract characteristics for the first formant of close front vowels like /IY/. With this model, the resonance frequency F_1 is related to the resonators parameters by

$$F_1 \propto \sqrt{A/(Vl)}, \tag{2.108}$$

where V is the volume of the back cavity, A is the cross-sectional area, and l is the length of the narrow tube. Fant (1973) proposed to model a change in the vocal tract size as a scaling of V and A by a factor k. With that the first formant would be scaled by $1/\sqrt{k}$ and, in contrast to the model of an open vowel as described above, a change in vocal tract length has a smaller effect on the first formant frequency.

Even though these differences in the spectral effects of VTL changes between different phonemes are well known, in practice the lossless, uniform tube model with the frequency relations given by Equation (2.107) is the basis for most implementations of *vocal tract length normalization* (VTLN) methods in ASR. VTLN tries to warp the filter bank center frequencies used within the feature extractor such that an uniform formant frequency scaling is approximated (Lee and Rose, 1996). Given the relation between formant frequencies and VTL as shown in Equation (2.107), a simple warping function for compensating this scaling effect is shown

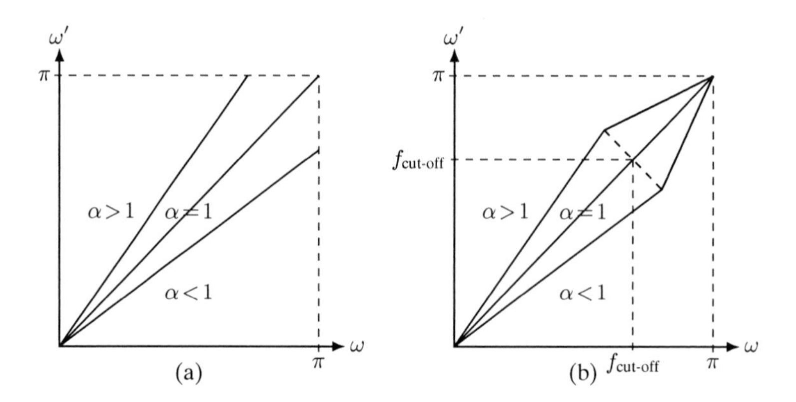

Figure 2.12.: (a) Linear warping function, (b) piecewise-linear warping function for VTLN (Hain et al., 1999)

in Figure 2.12 (a) with different warping factors α. In order to mitigate the problem of loosing information at high frequencies in case of extreme warping factors, different warping functions have been proposed. For example, Eide and Gish (1996) proposed an exponential warping function and Acero and Stern (1991) made use of a bilinear warping function. Experiments by Uebel and Woodland (1999) showed no significant differences in the resulting accuracies between different warping functions, which were also mathematically motivated.

Due to its simplicity, a piecewise-linear warping function (Hain et al., 1999) as illustrated in Figure 2.12 (b) is most often used for VTLN in practice. Within the decoding stage, a warping factor can be determined by a maximum-likelihood grid-search. Therefore, an utterance is decoded with a neutral warping factor of $\alpha = 1.0$ in a first decoding pass. Afterwards, forced alignments of the hypothesis are estimated for all considered warping factors. The factor that gives the highest likelihood is chosen as final warping factor, and a final decoding is done with this factor (Lee and Rose, 1996; Welling et al., 2002). Another approach for estimating the warping factor is the direct estimation of the formant frequencies (Eide and Gish, 1996). Alternatively, an optimal warping factor can be determined with a warping factor classifier. This approach is also known as *mixture-based warping factor estimation* (Lee and Rose, 1998). Once the warping factors for a given training set are determined, a classifier is trained that decides for the warping factor to be used for decoding the given unwarped utterance. This approach is computationally more efficient compared to the grid-search approach. However, because the temporal information is not used for the estimation of the warping factor, it is reported by Lee and Rose (1998) that the error rate slightly increases in comparison to the HMM-based warping factor estimation approach.

Besides its application in the decoding stage VTLN can also be used to obtain an acoustic model set that is less influenced by the speaker-induced variations. These speaker-independent models are often also referred to as *canonical* acoustic models. Training strategies were presented, for example, by Lee and Rose (1996) and Welling et al. (2002), both of which follow the same paradigm: Starting with a speaker-dependent acoustic model set, warping factors are estimated for each training speaker according to a maximum-likelihood criterion. Given the warping factors, a new model set is estimated based on the transformed features. This procedure is repeated for a certain number of iterations or until a maximum-likelihood criterion has converged. This kind of training is also known as *speaker-adaptive training* (SAT). Both, SAT and VTLN during decoding is explained in more detail in Section 6.1.1, where an enhanced VTLN approach is presented.

Besides the use of a single warping factor per utterance or per speaker, some works investigate the use of more than one warping factor. Maragakis and Potamianos (2008), for example, propose a region-based VTLN algorithm, where different warpings are applied on temporal regions. The regions were determined with a preceding classification step. In their work, an increase of word accuracy of up to ten percent relative could be observed. As a trade-off, however, the computational cost compared to the one-warping-factor VTLN approach as described above is further increased. A VTLN approach that estimates warping factors on a per-frame basis was proposed by Miguel et al. (2005). They augment the state-space of the Viterbi decoder with a locally constrained search for optimal warping factors. While this approach yields an even higher degree of freedom for VTLN compared to region-based VTLN, the computational costs are also higher. Furthermore, the warping-related decoder constraints and the more complex training procedure yield an additional amount of complexity to the ASR system.

While the VTLN methods that were described so far usually work with a linear frequency scale, it is also possible to perform a normalization by translating the spectral values on a logarithmic scale (Sinha and Umesh, 2002). Formally, for the magnitudes S of the STFT of two speakers A and B of the same vowel section the linear scaling on a linear frequency becomes a translation on a logarithmic frequency scale,

$$S_A(\omega) = S_B(\alpha \cdot \omega) \quad \Longleftrightarrow \quad S_A(\log \omega) = S_B(\log \alpha + \log \omega). \tag{2.109}$$

To distinguish between the VTLN approaches that perform a scaling of the linear frequency axis and those that perform a translation of the logarithmic frequency axis, these VTLN methods are referred to as *scaling VTLN* and *translational VTLN*, in the following. In practice, the quasi-logarithmic mel- or ERB-scale is used instead of the log-scale, because it was shown that these scales better map changes of the VTL to translations along the frequency axis (Umesh et al., 2002b). Additional experiments that confirm these findings have been conducted within this work and are described in detail in Section 4.3.

2.5.2. Maximum Likelihood Linear Regression

While vocal tract length normalization techniques commonly work within the feature space for normalization, there exist model-space techniques that adapt the parameters of the acoustic models to the characteristics of the observations from the current speaker. *Maximum likelihood linear regression* (MLLR) techniques represent the most prominent group of methods with this approach. There are two general forms of MLLR: unconstrained and constrained MLLR (Leggetter and Woodland, 1995; Gales, 1998). In *unconstrained* MLLR, separate linear transforms $\boldsymbol{A}$ and $\boldsymbol{H}$ are estimated that adapt the mean vectors and covariance matrices of component m of state j for each speaker $\mathtt{s}$,

$$\widehat{\boldsymbol{\mu}}_{\mathtt{s}}^{(jm)} = \boldsymbol{A}_{\mathtt{s}}\boldsymbol{\mu}^{(jm)} + \boldsymbol{b}_{\mathtt{s}}; \quad \widehat{\boldsymbol{\Sigma}}_{\mathtt{s}}^{(jm)} = \boldsymbol{H}_{\mathtt{s}}\boldsymbol{\Sigma}^{(jm)}\boldsymbol{H}_{\mathtt{s}}^{\mathrm{T}}. \tag{2.110}$$

In case of *constrained* MLLR (CMLLR), the same transform $\boldsymbol{A}$ is used for both, mean vectors and covariance matrices of the acoustic models,

$$\widehat{\boldsymbol{\mu}}_{\mathtt{s}}^{(jm)} = \boldsymbol{A}_{\mathtt{s}}\boldsymbol{\mu}^{(jm)} + \boldsymbol{b}_{\mathtt{s}}; \quad \widehat{\boldsymbol{\Sigma}}_{\mathtt{s}}^{(jm)} = \boldsymbol{A}_{\mathtt{s}}\boldsymbol{\Sigma}^{(jm)}\boldsymbol{A}_{\mathtt{s}}^{\mathrm{T}}. \tag{2.111}$$

The transforms are estimated such that the likelihood for the observations given the transformed acoustic models is maximized (Gales, 1998). An advantage of the MLLR approach is that it can adapt to the amount of available data: When only a small amount of data is available, a single transform can be estimated for all acoustic models. When more data is available, transforms for individual clusters of components can be estimated, which leads to a refined adaptation.

A *regression class tree* is used to determine the number of transforms for a given amount of data (Leggetter and Woodland, 1995). It is a binary tree and ties transforms to clusters of Gaussian components that are close in acoustic space. The tree is built by using the speaker-independent model set. The terminal nodes are denoted as base classes and determine the final component clusters. An exemplary regression class tree is shown in Figure 2.13. In this example, a regression class tree with four base classes (filled gray) is given. Solid lines indicate a sufficient amount of data for that particular class, while a dashed line indicates that there is insufficient amount of data for an estimation of a transform for that class. Here, classes 5, 6, and 7 would have an insufficient amount of data. However, by pooling

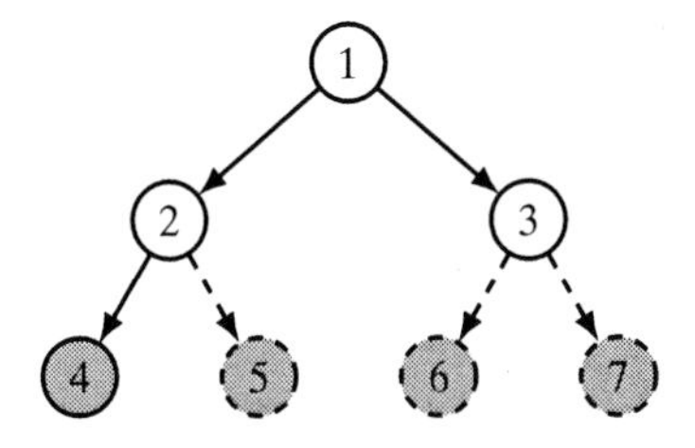

Figure 2.13.: An exemplary regression class tree with four base classes (gray). Adapted from (Young et al., 2009, p. 157).

classes 6 and 7 at node 3, there is a sufficient amount of data. The amount of data that is "sufficient" is given as a parameter prior to the construction of the regression class tree. Thus, transforms would be estimated for nodes 2, 3, and 4. The transform estimated for node 2 would be applied on all Gaussian components belonging to cluster 5, the transform from node 4 on the components belonging to cluster 4, and the transform estimated for node 3 would be applied on all Gaussian components that belong to clusters 6 and 7.

MLLR can be used in a supervised as well as in an unsupervised mode (Gales and Young, 2008). If no label information is available, a transcription is hypothesized with a speaker-independent model set. The hypothesis is then used for the estimation of the adaptation transform(s) in an iterative procedure until a convergence criterion is reached. Speaker-adaptive training with (C)MLLR can be carried out with a similar iterative procedure as SAT is done with VTLN (see Section 2.5.1 on page 52). Anastasakos et al. (1996) showed that the use of acoustic models as the result of SAT yields higher accuracies when using MLLR than adapting speaker-independent acoustic models that were not adaptively trained.

Different works have investigated the combination of VTLN and MLLR within a single ASR system. Pye and Woodland (1997) observed in their experiments that the beneficial effects of VTLN and MLLR are approximately additive. However, in Uebel and Woodland (1999) it is reported that this additive effect is not observable if multiple iterations of constrained MLLR were performed. This was investigated in detail by Pitz and Ney (2005), were it was formally shown that VTLN amounts to a special case of MLLR. Nonetheless, current state-of-the-art ASR systems usually make use of both, VTLN and (C)MLLR to achieve highest possible accuracies (see, for example, Stolcke et al., 2006; Sainath et al., 2010, 2011).

2.6. System Combination

A common way for further enhancing the performance of ASR systems is to combine the output of several systems that are used in parallel. The aim of system combination is to produce a hypothesis that is more accurate than those of the individual systems. The outputs to be used could be the actual hypotheses, the corresponding scores, or the output lattices. One form of system combination is *recognizer output voting error reduction* (ROVER) as proposed by Fiscus (1997). ROVER combines the 1-best hypotheses of each system in a two stage procedure: First, all hypotheses are aligned with each other by a dynamic programming procedure. In the second step, a voting scheme is used to decide for the best transcription. The scheme is parametrized such that at its extremes either a pure voting scheme according to the word frequencies or a decision based solely on the confidence scores is performed. An optimal value for this trade-off parameter is determined by minimizing the word error rate on a training set (Fiscus, 1997). In this work, the implementation available from Nat (2010) is used.

Instead of using only the 1-best hypothesis of each system, confusion networks (Mangu, 2000) based on the N-best hypotheses can be combined with each other (Evermann and Woodland, 2000). This approach is known as *confusion network combination* (CNC). However, no clear superiority of ROVER or CNC across varying numbers and types of ASR systems is observable (Evermann and Woodland, 2000; Hoffmeister et al., 2006).

3

Baseline Systems and Performances

The last chapter described the methods that are used within various stages of an ASR system as depicted in Figure 2.1. This chapter is about the practical parameter choices for the feature extraction schemes, for the acoustic modeling, as well as for the language modeling and normalization/adaptation methods. For conducting the experiments, a speech recognition framework based on the *hidden Markov model toolkit* (HTK, Young et al. (2009)) is used. The framework has the same architecture as described in Section 2.1. Furthermore, baseline performances of the standard feature types as described in the previous chapter are given in this chapter. While brief descriptions of the individual tasks are given together with the description of the baseline performances within this chapter, details about the corpora can be found in Appendix B.

3.1. Feature Extraction

All experiments used a window length of 20 ms and a window offset of 10 ms for the extraction of features from a speech signal. Depending on the recognition task, the sample rate of the input signals was either 8 or 16 kHz. Baseline error rates were computed for MFCC, PLP, and PNCC features. For each of these feature types 12 cepstral coefficients are extracted and supplied by the logarithmized energy. Furthermore, the corresponding delta and delta-delta features are concatenated to the feature vectors. In total, MFCC, PLP, and PNCC feature vectors are 39-dimensional. The considered temporal context of the delta and delta-delta features is 80 ms and 120 ms, respectively. With the given window length and offset, this is equivalent to seven and eleven frames, respectively.

MFCC In case of MFCC features, widely applied parameters (see, for example, Young et al., 2009) are used for their extraction: A mel filter bank with 24 triangular-shaped filters is used and covers the frequency range from 0 Hz to the Nyquist frequency. As described in Section 2.2.1, the filter center frequencies are equally spaced along the mel scale. The corresponding filter coefficients are normalized such that the coefficients of each filter sum up to one. The output values of the mel filter bank are compressed with the logarithm to the base 10. After computing the DCT coefficients for each frame, the first coefficient is replaced by the logarithmized energy feature. Furthermore, the component-wise mean is subtracted as a final step (see also "cepstral mean normalization" in Section 2.2.6).

PLP For the computation of PLP coefficients, the implementation from Ellis (2005) is used as basis. The number of critical-bands is set to 17 or 21 in case of a sampling rate of 8 kHz or 16 kHz, respectively. This corresponds to a filter bandwidth of one Bark per filter. The minimum frequency is set to 124 Hz and the Nyquist frequency is used as maximum frequency. For the LP analysis a model order of 12 is used. If RASTA processing is enabled for the PLP features, the implementation that is provided by Ellis (2005) is used. The PLP coefficients were transformed to cepstral domain and supplied by the log-energy feature. Also, delta and delta-delta features were concatenated. Overall, the PLP feature vector consisted of 39 components.

PNCC The computation of PNCC features is based on an adapted implementation provided by Kim and Stern (2010a). For reasons of efficiency the gammatone filter bank that is based on the implementation of Slaney (2011) was used. The smoothing and nonlinearity parameters as well as the "delta" parameter of the power bias subtraction algorithm (see Section 2.2.6 on page 29) are chosen as described by Kim and Stern (2010a). As done with the MFCC and PLP features, also the PNCC features were amended with the log-energy and delta features. The final dimensionality of the PNCC features was also 39.

3.2. Acoustic Models

For small vocabulary recognition tasks like AURORA or OLLO (see Appendix B) whole-word models were used as acoustic models. The TIMIT corpus was specifically designed for acoustic-phonetic studies and speech recognition experiments with emphasis on the phonemic level. Therefore, phonemes were used as acoustic models and phoneme recognition experiments were conducted on this corpus. The number of Gaussians for the GMMs that were used by the acoustics models was set to 16 in case of TIMIT, 8 in case of OLLO, and 4 in case of TIDIGITS and AURORA. The covariance matrices were constrained to be of diagonal form.

3.3. Language Models

In case of TIDIGITS, AURORA, and OLLO, simple word networks as described in Young et al. (2009, p. 187) are used for defining the allowed word sequences. In case of OLLO, for example, the allowed sequences consisted of a silence model, followed by one of the 150 logatomes, followed by the silence model. The experiments on the TIMIT corpus make use of a bigram language model that is generated on base of the provided training data set of TIMIT.

3.4. Parameter Estimation

The parameter estimation procedures for the individual tasks follow the one depicted in Section 2.4.2 with triphone models at the finest modeling level. Some of the details about the training are given in the following: After a flat-start initialization, single Gaussian monophone models are trained with four iterations of Baum-Welch training. If enabled, a short-pause model is created afterwards. Based on the monophone models all triphone models that are contained in the training set are then constructed. After another iteration of Baum-Welch training the parameters of the triphone models are tied by performing a rule-based clustering with the question set that was proposed by Vertanen (2011). This is again followed by a four iterations of the Baum-Welch algorithm. The number of Gaussians for the probability density functions is set to be proportional to the available amount of training data for the individual models. In case of the considered feature types MFCC, PLP, and PNCC a maximum of 16 Gaussians mixtures is used. To speed up the whole training procedure, the parallelization capability of the HTK tool to reestimate the parameters of the acoustic models is used. Further details for the parameter estimation with HTK can be found in the descriptions from Young et al. (2009) and Vertanen (2011).

3.5. Normalization and Adaptation

Normalization and adaptation methods were implemented for the ASR framework for speaker-adaptive training (SAT), as well as for the decoding stage. For VTLN, the procedures that were described by Welling et al. (2002) were implemented. The maximum-likelihood grid-search for optimal warping parameters considers a set between 13 to 17 warping parameters. For a (piecewise-)linearly scaled frequency axis, a range from 0.88 to 1.12 is used for the warping values. This is a commonly used range for VTLN (see, for example, Lee and Rose, 1998; Welling et al., 2002; Maragakis and Potamianos, 2008). For a given warping parameter α, the log-likelihood of the original and the transformed feature vectors $\boldsymbol{y}$ and $\boldsymbol{y}^{\alpha}$, respectively, is given by

$$\log(p(\boldsymbol{y};\, \alpha, \boldsymbol{\lambda})) = \log(p(\boldsymbol{y}^{\alpha};\, \boldsymbol{\lambda})) + \log(J^{\alpha}), \tag{3.1}$$

where J^{α} is the Jacobian of the implicit VTLN transformation for warping parameter α (Kim et al., 2004). In case of a linear transformation $\boldsymbol{A}^{\alpha}$ as warping function, the Jacobian is given by $J^{\alpha} = |\boldsymbol{A}^{\alpha}|$. To mitigate the effect of the Jacobian term in Equation (3.1), the mean and the variances of the feature vector components are normalized. By doing so, the Jacobian J^{α} is approximately the same for each warping factor (Kim et al., 2004). This allows for a comparison of the log-likelihoods during the grid-search for an optimal warping parameter.

If MLLR is used in the experiment, SAT is performed after the estimation of the tied-state triphone models (see also Section 3.4). A regression class tree with eight base classes is computed on base of the triphone parameters. The corresponding components of the silence and short-pause models are contained within an explicit nonspeech leaf of the tree. If dynamic features are contained within the feature vectors, the MLLR transformations can be constrained to a block-diagonal form, which decreases the number of parameters to be estimated. This has been beneficial to the systems performance throughout the experiments of this work. For SAT, CM-LLR transformations were estimated for each speaker, while for the decoding stage CMLLR and MLLR transformations were estimated. In practice, the estimation of the speaker-dependent transformations was done in parallel.

3.6. Decoding

In this work the Viterbi decoder "HVite" from HTK (Young et al., 2009) was used.A beam search was not used during the experiments. If VTLN and/or MLLR were enabled, a two-pass approach is used. First, a hypothesis based on nontransformed features is generated. Then, the hypothesis is used to estimate the most likely warping parameter or the (C)MLLR transformations, respectively. The found warping parameter/transformation is then used to generate another hypothetical transcription that is used in a second turn to re-estimate the most likely warping parameter/transformations. The ASR framework is able to perform the decoding of several utterances in parallel.

3.7. Baseline Performances

In the following, baseline performances for various tasks and feature types (that are described in Section 2.2) are presented. While a brief description of the individual corpora is given here, more detailed descriptions of the corpora can be found in Appendix B. A comparison of accuracies as result from different feature extraction method can only be valid, if the same training and test data is used, and also, if the acoustic and language modeling of the ASR system are the same. For convenience, a list of comparable results (with respect to the task, the acoustic modeling, and the language modeling) is shown for each corpus in Appendix B.

TIMIT The TIMIT corpus (Garofolo et al., 1993) is a phonetically rich and hand-labeled small vocabulary corpus with around five hours of training and test utterances. Though, real-world ASR systems are trained on hundreds of hours of speech, Sainath et al. (2009) points out that TIMIT still plays an important role in the algorithmic development, because gains observed on this task can also be observed on larger corpora. Because of its size TIMIT was the primary corpus during the development of many of the methods presented within this work. The experiments conducted on the TIMIT corpus are phoneme recognition experiments. To allow for the comparison with other research groups, a de facto standard procedure for conducting phoneme recognition experiments on TIMIT was established by Lee and Hon (1989): For acoustic modeling, the original phoneme set of TIMIT, which consists of 61 phonemes, is reduced to 48 phonemes. After recognition the phoneme set is further reduced to 39 phonemes for performance evaluation. In the following, baseline accuracies are given for two scenarios that imitate matching mean VTLs and mismatching mean VTLs within the training and the test sets.

Table 3.1.: Baseline accuracies [%] for the experiments on the TIMIT corpus with and without MLLR. Accuracies are shown for matching and mismatching training-test conditions.

	scenario & adaptation			
	matching		**mismatching**	
Feature type	**-**	**MLLR**	**-**	**MLLR**
MFCC	72.2	75.2	54.7	66.9
PLP	72.3	75.4	54.3	66.4
RASTA-PLP	68.3	70.8	48.4	58.1
PNCC	70.5	72.9	50.7	62.3

Table 3.1 shows the baseline accuracies for MFCC, PLP, RASTA-PLP, and PNCC features for TIMIT.

The table shows the accuracies for each feature type for the cases when no MLLR is used, and for the case when MLLR is used. The accuracies for the matching training-test scenario are given in the first and second columns, and the accuracies for the mismatching scenario are given in the third and fourth columns. Without MLLR it can be seen that MFCC and PLP features perform best in comparison to RASTA-PLP and PNCC features in case of matching conditions. This holds for the absolute accuracies, as well as for the decreases in accuracy in case of mismatching training-test conditions. The lower performance of RASTA-PLP and PNCC features under clean-speech conditions in comparison to MFCC and PLP features can be seen as a trade-off for a higher noise robustness. This can also be observed in the experiments on the AURORA 2 task within this work (see next paragraph). In case of mismatching training-test conditions the accuracy of each feature type decreases by 17 to 20 percentage points compared to their corresponding accuracy with matching training-test conditions. The sensitivity of the standard feature types to VTL differences can clearly be observed here. When using MLLR it can be observed that this decrease in accuracy is smaller (between about eight and twelve percentage points) for all feature types.

AURORA 2 The AURORA 2 task is used within this work to investigate the robustness to additive noise of the different feature types. As described in more detail in Appendix B.2, AURORA[1] provides two predefined ways for training:

[1] There also exists a corpurs called AURORA 4. Since AURORA 4 is not used in this work, the notion of AURORA referrs to the AURORA 2 corpus in this work.

Table 3.2.: Baseline word recognition accuracies [%] for the experiments on the AURORA 2 corpus for (a) clean-speech training and (b) multi-style training. The accuracies for SPLICE-MFCC are taken from the work of Droppo et al. (2001).

	SNR [dB]						
Feature type	**∞**	**20**	**15**	**10**	**5**	**0**	**-5**
MFCC	98.6	96.8	93.0	78.1	51.2	26.3	12.2
PLP	98.4	96.1	88.7	67.1	43.4	21.8	10.6
RASTA-PLP	98.4	96.3	94.0	88.6	75.8	50.7	23.5
PNCC	98.6	97.7	95.7	90.1	75.7	49.1	21.9
SPLICE-MFCC	99.1	98.2	97.0	93.1	81.6	55.1	22.4

(a)

	SNR [dB]						
Feature type	**∞**	**20**	**15**	**10**	**5**	**0**	**-5**
MFCC	98.4	97.9	96.8	93.9	85.3	65.1	31.1
PLP	97.9	97.4	96.3	93.7	85.0	63.2	25.6
RASTA-PLP	97.0	97.2	96.4	94.3	87.5	70.2	37.7
PNCC	98.0	97.7	97.0	95.1	88.3	72.1	40.2
SPLICE-MFCC	98.9	98.3	97.2	94.4	86.0	62.9	26.9

(b)

either clean-speech or multi-style training. The accuracies of the standard features for both training styles are shown in Table 3.2, where (a) refers to the clean-speech training style and (b) refers to the multi-style training style. The feature type SPLICE-MFCC refers to SPLICE-enhanced MFCCs as described in Section 2.2.6. The results for this feature type are taken from the work of Droppo et al. (2001). Because a slightly different setup was used in case of SPLICE-MFCCs, the comparability of the accuracies for SPLICE-MFCCs is limited. However, these results were incorporated into the table to show results for a supervised feature-enhancement method.

Looking at the accuracies in case of clean-speech training, it can be observed that all feature types have a similar performance in case of clean speech test data. However, with a decreasing SNR the two feature types RASTA-PLP and PNCC, which were both specifically designed for noisy conditions, perform significantly better than MFCC and PLP features. With an SNR of minus five decibel the accuracy of

Table 3.3.: Baseline accuracies [%] for the experiments on the OLLO corpus.

	speaking rate		
Feature type	**slow**	**normal**	**fast**
MFCC	55.7	63.5	56.0
PLP	60.6	67.0	55.8
RASTA-PLP	50.4	64.2	49.7
PNCC	58.0	66.1	53.0

RASTA-PLP and PNCC is about ten percentage points higher compared to MFCC and PLP features. Comparing RASTA-PLP and PNCC features, a slightly higher accuracy with PNCC features can be observed. When the acoustic models are trained on multi-style data, a slight decrease in accuracy in case of clean speech can be observed. This can be explained by the slight mismatch between the training data and the clean-speech test data. As expected, all four feature types achieve much higher accuracies for lower SNRs than with clean-speech training. Similar to the clean-speech training, PNCC features perform best under all noise conditions.

OLLO The OLLO corpus (Wesker et al., 2005) has been designed especially for experiments about intrinsic speaker variabilities. Among others, one of these variabilities is the rate of speech. The different speaking rates of the utterances are grouped into the categories "slow", "normal", and "fast". Even though, the rate of speech is not in the main focus of this work, experiments were conducted that evaluate the different feature types on this task. This gives further insights and might motivate future works. Overall, 150 different *logatomes* are present within the data set. Logatomes are sequences of consonant-vowel-consonant or vowel-consonant-vowel, where the first and the last phoneme are the same. For the experiments in this work only the utterances that belong to one of these categories were used. The logatomes were mapped to phoneme sequences according to the *Carnegie Mellon University Pronunciation Dictionary* (CMUDict) (CMU, 2011). With that mapping 25 phonemes were modeled with three-state left-to-right HMMs without any state skips. Since only a single logatome is contained within a single utterance, a language model with equal priors for each logatome is used. For training only the utterances that belong to the group of "normal" speaking rates were used. For testing the utterances from all three groups were used. The recognition rate was determined according to the number of correctly recognized center phonemes of the logatomes. Table 3.3 shows accuracies for MFCC, PLP, and PNCC features for the experiments that were conducted on the OLLO corpus.

In case of a low and normal speaking rate it can be seen that PLP features yield the highest accuracy in comparison to the other three feature types. In case of a high speaking rate, MFCC and PLP features perform similar, while the RASTA-PLP and PNCC features, which are specifically designed for noisy conditions, perform worse. The accuracies for the utterances that belong to the "fast" group are lower than the accuracies for the utterances belonging to the "slow" group. This was also observed in the work of Meyer et al. (2010).

3.8. Choice of Filter Bank

Even though, the mel filter bank models the time-frequency analysis of the auditory system only rudimentary, their efficient computation and their resulting degree of phoneme separability make them the first choice for many applications. However, as described in more detail in Section 2.2, there exist other types of filter banks that better explain psychoacoustic observations. Instead of computing the logarithm of the spectral values (as in case of MFCC features), the filter bank outputs can be passed to a nonlinear power-function to compress the dynamic range of the spectral values. The extraction of PLP coefficients, for example, includes a power-function with exponent 0.33, and for the PNCC features an optimal choice for the exponent was reported to be 0.1 (Kim and Stern, 2009) and in a later work 0.15 (Kim and Stern, 2010a). From a physiological point of view, a compression of the values at this stage imitates the power law of hearing and simulates the nonlinear relation between the intensity of a signal and its perceived loudness (Hermansky, 1990). From an acoustic modeling point of view, different power-functions lead to different PDFs of the features, which allow for a better or worse modeling with a fixed number of Gaussians.

In the following, the different types of filter banks that were described in Section 2.2.1 are compared with respect to their performance in an ASR task. The results of these experiments motivate the later choice of filter bank for the investigated invariant feature extraction methods that are explained in detail in Chapter 4. For the filter bank comparison the mel, gammatone, static gammachirp, and dynamic-compressive gammachirp filter banks were considered. The TIMIT corpus with its standard training and test set (see Appendix B.1) was chosen as database. Triphones were used as acoustic models together with a bigram language model. No VTLN or MLLR was applied during training or testing. This was done to decrease the number of involved methods within the ASR system so that the impact of the different filter banks can be better evaluated. The insertion penalty and grammar scaling factors were tuned in each experiment. Similar to the common MFCC setup (for example, Young et al., 2009, details in Section 2.2), 24 channels were used

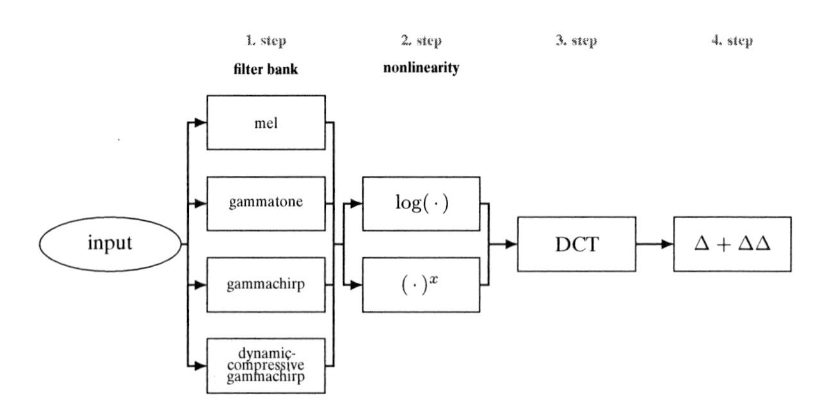

Figure 3.1.: Overview of the processing schemes that were considered in the experiments about the different choices for a filter bank and nonlinearity. Different types of filter banks and different choices for the nonlinearity were considered in the experiments.

Table 3.4.: Accuracies for $\log_{10}$-based cepstral coefficient features for different types of filter banks on the TIMIT corpus.

Filter bank	**Accuracy [%]**
mel	73.4
gammatone	73.8
gammachirp (static)	72.4
gammachirp (dynamic)	70.2

in all filter banks. The lowest frequency was set to 40 Hz, the highest frequency was set to 8 kHz. Depending on the type of filter bank the center frequencies of the filters were evenly spaced on the mel or the ERB scale. In a first experiment, the outputs of the filter banks were passed to a logarithm nonlinearity as it is done for the computation of MFCC features. The first 13 DCT coefficients (including the DC coefficient) of each frame were computed and supplied with delta and delta-delta features. In a second experiment, the logarithm operator was replaced by a nonlinear power-function with exponents ranging from 0.02 to 0.3 in steps of 0.02. Figure 3.1 shows an overview of the considered processing schemes. The resulting accuracies of this experiment are summarized in Table 3.4.

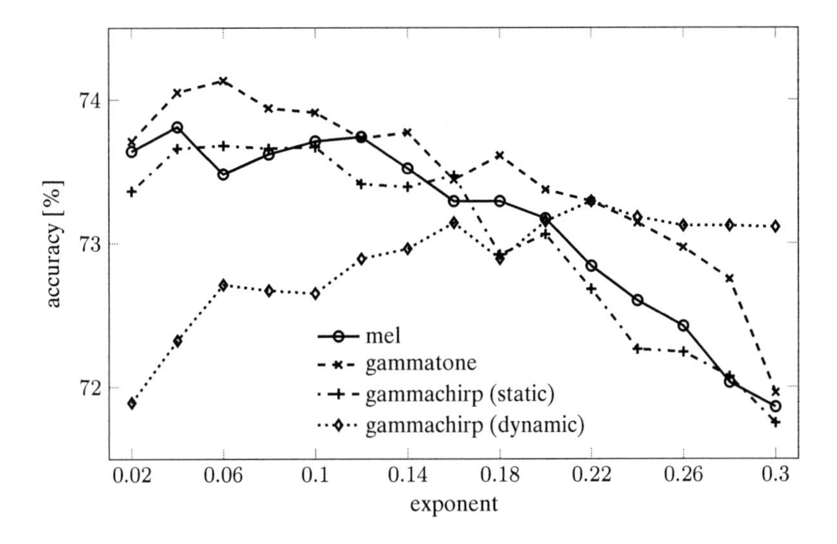

Figure 3.2.: Accuracies for cepstral coefficients with power-law compressed spectral values for different types of filter banks on the TIMIT corpus.

It can be seen that the cepstral features based on the mel and the gammatone filter bank perform similar. This similarity in recognition performance was also reported by Schlüter et al. (2007). Both the static and the dynamic-compressive gammachirp filter bank, yield accuracies that are lower than the cases in which a mel or gammatone filter bank is used. This is somewhat surprising since the static as well as the dynamic-compressive gammachirp filter bank have been designed to explain psychoacoustic observations better than the gammatone filter bank. The parameters of the implementation of the gammachirp filter bank were used as provided by Irino. An optimization of these parameters might yield an improved performance within these experiments.

As mentioned above, the logarithm operator was replaced by a nonlinear power-function with exponents ranging from 0.02 to 0.3 in steps of 0.02 in a second experiment. The results of this experiment are shown in Figure 3.2. It can be observed that different choices for the exponent of the power-function have a significant impact on the performance of the recognition system. Overall, the accuracies range from 71.7% to 74.2%. With respect to the highest accuracies for the individual types of filter bank it can be seen that the gammatone filter bank performs best. In comparison with each other, the mel and the static gammachirp filter banks perform similar well. The dynamic-compressive gammachirp filter bank yields the lowest accuracies in this experiment. Walters (2011, p. 164f) also reports that the use of a dynamic-compressive gammachirp filter bank lead to worse accuracies than a gammatone filter bank. In that work it is supposed that a tuning of the internal parameters of the filter bank, as well as of the subsequent feature extraction method might increase the performance of the ASR system. The results of the experiments in this section also allow for the following conclusions: Even though, the dcGC filter bank better explains observations that were made in psychoacoustic experiments, a direct replacement of a mel or gammatone filter bank in an ASR front-end is not beneficial without further parameter tuning. The results of these experiments motivate the use of the gammatone filter bank for the time-frequency analysis stage of the invariant feature extraction methods described in the next chapter.

4

Invariance Transforms for Feature Extraction in ASR

In the first part of this chapter the principles of invariant feature extraction methods are described. An overview of related works of invariant extraction methods for ASR is given in Section 4.2. A central element of within the theory of invariants is the transformation group for which invariance should be achieved. As will be described in more detail in the Section 4.3, the group of translations can be related to different VTLs in the field of ASR and, hence, is the transformation group for which invariance should be achieved. The last three section of this chapter describe feature extraction methods that make use of different translation-invariant transforms. In chapter 4.4 transforms of the class $\mathbb{C}T$ are used for feature extraction. In Section 4.5 a feature extraction method that makes use of the "generalized cyclic transform" is described. While transforms of the class $\mathbb{C}T$ and of the class GCT where specifically designed for translation-invariance, there exist another method known as "invariant integration". This is a constructive approach to obtain invariance against arbitrary transformation groups. A feature extraction method for speaker-independent ASR that makes use of invariant integration is described in the last section of this chapter.

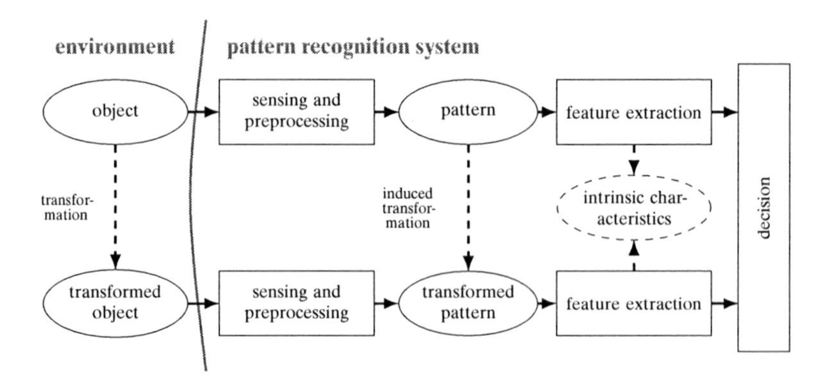

Figure 4.1.: Relation between transformed real-world objects and their parametric machine representations. Adapted from Schulz-Mirbach (1995a).

4.1. ASR as Pattern Recognition Problem and Principles of Invariance Transforms

Generally speaking, a *pattern recognition system* gets sensed data from a real world object as input and groups this data into classes, which are identified by class labels. Such a system can be decomposed into several components. These are depicted in the top row of Figure 4.1. From the point of view of an ASR system, the real world object is, for example, the uttered sound wave of a speaker. The environment of the speaker has an influence on the object and might be described, for example, by the acoustic conditions around the ASR sound sensor or by the characteristics of the sensor itself. The sound wave is sensed by one or more microphones and converted into a digital signal. An exemplary preprocessing step might be a mean normalization of the input signal. At this stage, the sound wave is represented within the ASR system as a digitized speech signal with certain patterns in the time- and in the corresponding frequency-domain. These patterns are used to extract parameters, which optimally are very similar for objects of the same category, and which allow for a separation between objects of different classes. These parameters are commonly referred to as *features* (Duda et al., 2001). The feature extraction methods considered in this work are described in Section 2.2. The decision stage of the pattern recognition system uses the extracted features to generate a mapping to one ore more class labels. In case of ASR systems, the common decoding procedure is described in Section 2.4.

A general problem of pattern recognition systems is a large variability of the real world objects, which might lead to dissimilarities between features of objects of the same class if not specifically handled. Different kinds of variabilities within the field of ASR are described in Section 1.2. A variability with a major focus within this work is the vocal tract length (see Section 1.1 on page 2ff). Generally, the different realizations of real-world objects that belong to the same class can be related by certain *transformations* within the environment. These transformations *induce transformations* within the pattern space of the pattern recognition system. Ideally, a feature extraction method yields the same parameter values for every (transformed) pattern of the same class. In that case, a feature describes an *intrinsic characteristic* of a pattern. Features with such a property are referred to as *invariant* features. Nonlinear transformations that lead to invariant features have been investigated and successfully applied for decades in the field of pattern recognition. A brief introduction to the general notions and concepts for the construction of invariants is given in the following. It is based on the book chapter by Burkhardt and Siggelkow (2001) and the thesis of Schulz-Mirbach (1995a).

The idea of invariant features is to find a mapping T that is able to extract features that are the same for different observations $\boldsymbol{y}$ of the same equivalence class with respect to a group action G. Such a transformation T maps all observations of an equivalence class into one point of the feature space:

$$\boldsymbol{y}_1 \overset{G}{\sim} \boldsymbol{y}_2 \Rightarrow T(\boldsymbol{y}_1) = T(\boldsymbol{y}_2). \tag{4.1}$$

In our case, it means that all frequency-warped versions of the same utterance should result in the same sequence of feature vectors. Given a transformation T and an observation $\boldsymbol{y}$, the set of all observations that are mapped into one point in the feature space is denoted as the *set of invariants* $\mathcal{I}_T(\boldsymbol{y})$ of an observation:

$$\mathcal{I}_T(\boldsymbol{y}) = \{\, \boldsymbol{y}_i \,|\, T(\boldsymbol{y}_i) = T(\boldsymbol{y}) \,\}. \tag{4.2}$$

The set of all possible observations within one equivalence class is called *orbit* $\mathcal{O}(\boldsymbol{y})$: Given a prototype $\boldsymbol{y}$, all other equivalent observations can be generated by applying the group action G,

$$\mathcal{O}(\boldsymbol{y}) := \{\, \boldsymbol{y}_i \,|\, \boldsymbol{y}_i \overset{G}{\sim} \boldsymbol{y} \,\}. \tag{4.3}$$

A transformation $T$is said to be *complete* if both the set of invariants of an observation and the orbit of the same observation are equal. Complete transformations have no ambiguities regarding the class discrimination. Incomplete transformations, on the other hand, have the property

$$\mathcal{O}(\boldsymbol{y}) \subseteq \mathcal{I}_T(\boldsymbol{y}) \tag{4.4}$$

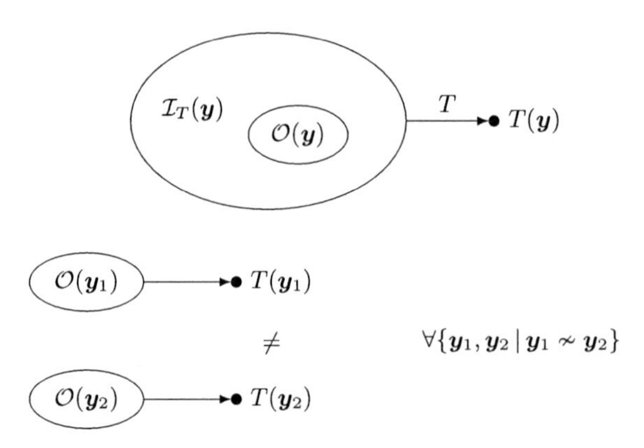

Figure 4.2.: (a) Relation of invariant set $\mathcal{I}_T$ and orbit $\mathcal{O}$ for a given transformation T and observation $\boldsymbol{y}$, (b) notion of *completeness* of a transformation T. Adapted from Burkhardt and Siggelkow (2001).

and may lead to the same features from observations of different equivalence classes and thus cannot distinguish them (Burkhardt and Siggelkow, 2001). The described terms are visualized in Figure 4.2.

In practice, a "high degree of completeness" is desired, which means that $\mathcal{I}_T(\boldsymbol{y})$ should not be much larger than $\mathcal{O}(\boldsymbol{y})$. Principles to systematically construct invariants with this property can be divided into three categories:

1. *Normalization.* Here, the observations are transformed with respect to extreme points on the orbit. Usually, a certain subset of the observations is chosen for the parameter estimation of the transformation.

2. *Differential approach.* This group of techniques obtains invariant features by solving partial differential equations. Its main idea is that the features should be insensitive to infinitesimally small variations of the parameter(s) of the group action. Let these parameters be denoted by $\boldsymbol{\beta} = (\beta_1, \beta_2, \dots, \beta_N)$. For a group element $g \in G$ and a term $g(\boldsymbol{\beta})$ denoting a transformation $g \in G$ with parameters $\boldsymbol{\beta}$ and an input signal $\boldsymbol{y}$, it is demanded that

$$\frac{\partial T(g(\boldsymbol{\beta})\boldsymbol{y})}{\partial \beta_i} \equiv 0, \quad i = 1, 2, \dots, N. \tag{4.5}$$

The solutions of the partial differential equations are the invariants.

3. *Integral approach.* The idea of the third group of methods is to compute averages of arbitrary functions on the entire orbit. Hurwitz invented the principle of integrating over the transformation group for constructing invariant features in 1897 (Hurwitz, 1897). The resulting integral is independent of the parameter(s) of the group action G:

$$T_f(\boldsymbol{y}) = \frac{1}{|G|} \int_G f(g\boldsymbol{y}) \mathrm{d}g, \tag{4.6}$$

   where $|G| := \int_G \mathrm{d}g$ is the "power"of the group, and f is a (possibly complex-valued) kernel function.

One drawback of the normalization approach is the problem of a proper choice of a subset of the parameters. Another disadvantage is that a reduction of dimensionality is not given with this method. For the differential approach, solving the partial differential equations is the main difficulty in practice. In contrast, the integral approach does not need any preprocessing and a method for feature extraction in ASR is described in detail in Section 4.6. Successful applications of the integration approach in the field of image processing were described, for example, by Schulz-Mirbach (1995b) and Siggelkow (2002).

The three described principles represent general approaches for the construction of invariants for arbitrary transformation groups with a high degree of completeness. Certainly, there exist other methods that are invariant to specific transformations. However, the set of invariants $\mathcal{I}_T(\boldsymbol{y})$ of these methods is generally much larger than the orbit $\mathcal{O}(\boldsymbol{y})$. This means that the generated features are invariant to more operations than desired. Examples of translation-invariant transformations are the magnitude of the Fourier transformation as well as the auto- and cross correlation functions. Both the Fourier-transform magnitude and the autocorrelation of a signal remain unchanged when applied to translated versions of the same signal. For the cross-correlation of two signals, both signals may be translated by the same arbitrary amount without affecting the result. Known invariance transforms from the field of image analysis are the *generalized cyclic transforms* (Lohweg et al., 2004) and the *transformations of the class* $\mathbb{C}T$ (Burkhardt and Müller, 1980), which include the *rapid transform* (Reitboeck and Brody, 1969). Although, theoretically, these transforms have a larger set of invariants compared to the methods of the described three groups, all of them proved to be valuable in different kinds of applications (Fang and Häusler, 1989; Lohweg et al., 2004). It is shown within this chapter, how the transformations of the class $\mathbb{C}T$ and the generalized cyclic transforms can be used for feature extraction in ASR.

The use of invariance transforms for ASR systems is a major focus of this work. In the following, an overview of related works on this field is given. Afterwards, three kinds of transforms are described and feature extraction strategies for ASR systems are presented and discussed.

4.2. Related Works about Invariance Transforms for ASR

The concept of extracting features for ASR that are independent of the VTL has been taken up by several works in the past, and different methods were proposed. In the following, a brief summary of these ideas is presented.

To begin with, Cohen (1993) introduced the *scale transform*, which, for a given function $X(f)$, is given by

$$D_X(c) = \int_0^\infty X(f) \frac{e^{-j2\pi \ln f}}{\sqrt{f}} \, \mathrm{d}f. \tag{4.7}$$

Cohen (1993) showed, that the scale transform is a special case of the *Mellin transform*. In analogy to the Fourier transform, Cohen (1993) refers to $|D_X(c)|^2$ as the *energy density scale spectrum*. One property of the scale transform is that the magnitude of the scale transform of a function $f(t)$ and the magnitude of the scale transform of a scaled version of this function $\sqrt{a}f(at)$ are the same. For $\sqrt{\alpha}X(\alpha f)$, the scale transform is given by

$$D_X^\alpha(c) = \int_0^\infty \sqrt{\alpha} X(\alpha f) \frac{e^{-j2\pi c \ln f}}{\sqrt{f}} \, \mathrm{d}f. \tag{4.8}$$

By substituting αf by f', it is

$$D_X^\alpha(c) = e^{+j2\pi c \ln \alpha} \int_0^\infty X(f') \frac{e^{-j2\pi c \ln f'}}{\sqrt{f'}} \, \mathrm{d}f', \tag{4.9}$$

$$= e^{+j2\pi c \ln \alpha} D_X(c). \tag{4.10}$$

Thus, the magnitudes can be seen to be scale invariant. Based on the assumption that the resonance frequencies between speakers with different VTLs are related by a linear scaling (see Equation (2.107) on page 53), the scale transform was investigated for its applicability in the field of ASR by Umesh et al. (1999a), where a *scale cepstrum* is obtained by computing the scale transform of the magnitude spectrum of a Fourier transformed segment of a speech signal. Umesh et al. (1999a) compared the LDA-based separability criterion (see Equation (2.66) on page 37) of MFCC features and the scale cepstrum of four vowels. It was shown that the scale-cepstral coefficients provide a higher separability than the mel-based cepstral coefficients. However, no recognition experiments were conducted in that work.

Also Sena and Rocchesso (2005) studied the scale transform for vowel recognition without conducting any recognition experiments on standard corpora like TIMIT or AURORA. Vowel classification experiments were presented by Umesh et al. (2007), where the scale transform was used to compute the scale cepstral coefficients as proposed by Umesh et al. (1999a). In that work it was shown that the *scale transform cepstral coefficients* (STCC) perform worse in a phoneme recognition task than MFCC features. Umesh et. al. suggest to use vowel-dependent and averaged phase-information to increase the performance of the STCC features. However, in that work the knowledge about the true transcription is assumed to be available in order to obtain results that perform better than the standard VTLN approach in combination with MFCC features. This makes their approach impractical in practice.

Patterson (2000) describes an auditory model of human speech perception which is referred to as *auditory image model* (AIM). Within the AIM, a segment of a speech signal is represented within a two-dimensional space denoted as *stabilized auditory image* (SAI). The Mellin transform was applied on the SAI in the work presented by Irino and Patterson (2002) in order to achieve a VTL normalization. A small number of synthetic vowels was used to demonstrate the VTL normalization capability of the Mellin transform in combination with the AIM. The effects of different VTLs within the SAI space have been analyzed in more detail in the work of Patterson et al. (2007), where it is shown that the space of the SAI is scale covariant. This means that a change of the VTL leads to a shift along the subband axis, as well as to a scaling along the time-interval axis. The scaling is caused by the different lengths of the impulse responses of the filters of the preceding filter bank. Walters (2011) presents detailed studies about the use of the AIM for syllable recognition tasks with GMM-parameter based features. Also in that work, Walters explicitly motivates the use of translation-invariant transforms on base of the later processing stages of the AIM for future work.

Various works about feature extraction for ASR rely on the assumption that the effects of inter-speaker variability caused by VTL differences is mapped to translations along the subband-index space of an appropriate filter bank analysis as described by Equation (2.109) on page 55. While Sinha and Umesh (2002) use this approximation for a VTLN based on translation, Mertins and Rademacher (2005, 2006) and Rademacher et al. (2006) present various translation-invariant feature extraction methods for ASR: Mertins and Rademacher (2005) used a wavelet transformation for the time-frequency analysis and proposed so-called *vocal tract length invariant* (VTLI) features based on auto- and cross-correlations of wavelet coefficients. These are based on the logarithmized correlation sequences $\log r_{ss}$ of spectral profiles $s_n(k)$,

$$\log r_{ss}(n, d, m) = \log \sum_k s_n(k) s_{n-d}(k+m), \quad (4.11)$$

and on the correlation sequences c_{ss} of logarithmized spectral profiles,

$$c_{ss}(n, d, m) = \sum_k \log(s_n(k)) \cdot \log(s_{n-d}(k+m)). \quad (4.12)$$

Subsequent work (Rademacher et al., 2006) showed that a gammatone filter bank instead of a previously used wavelet filter bank leads to a higher robustness against VTL changes. Furthermore, it was shown that the incorporation of phase information from a complex-valued TF represention into the feature vectors is benefitial to the overall performance of the ASR system. Qiao et al. (2009) describes *affine-invariant features* (AIF) to achieve an enhanced robustness to the effects of different VTLs. On a small-vocabulary isolated word recognition task, they showed that the AIFs are able to achieve higher accuracies than MFCC features. However, experiments on larger corpora and on continuous speech recognition tasks were not conducted.

In the next sections, three different methods for computing translation-invariant features are described. Each section first introduces the principles of each transform and describes ways of applying the individual transforms for extracting features for ASR. During the description of their individual methods, results that were published at the time of the development are described. The speech recognition systems used during the development of the individual methods are based different setups. To allow for a comparison of the individual feature types with each other, experiments with a unified speech recognition framework are described in Section 4.7. These experiments allow for a comparison with the baseline accuracies described in Section 3.7. Furthermore, the performance of system combinations of the individual ASR systems are described at the end of Chapter 4. To clarify which results within this work are comparable with each other, each corpus description in Appendix B contains a list of tables with comparable accuracies. Before the feature extraction methods are described, however, the use of a gammatone filter bank throughout this work is motivated by the results of the experiments that are described next.

4.3. Frequency Scale Optimization for Translational VTLN

As described in Section 2.2.1 different scales are commonly used for determining the center frequencies of the filters used within the feature extraction methods. These scales are based on findings from different psychoacoustic experiments (Stevens and Volkmann, 1936; Glasberg and Moore, 1990). According to the model of the vocal tract as a lossless, uniform tube (see Section 2.5.1), resonance frequencies are linearly dependent on the length of the vocal tract and, formally, a logarithmic frequency scale maps the effects of different VTLs to translations along the (logarithmic) frequency axis. However, a lossless, uniform model can only be a rough approximation and other scales might be more appropriate to map the spectral effects due to different VTLs to translations along the subband-index axis. On the one hand, results from the the work of Umesh et al. (2002a), for example, show that the quasi-logarithmic mel scale better maps the effects of different VTLs to translations. On the other hand, several authors (for example, Irino and Patterson, 1999; Sinha and Umesh, 2002; Monaghan et al., 2008) presented results that suggest that the ERB scale is better suited for such a mapping than the log or the mel scale. As is described in more detail in the following sections, the assumption that different VTLs lead to translations along the subband-index axis is fundamental. This section investigates the impact of various scales on the recognition rate of an ASR system with translational VTLN.

In the following, an experiment is described that investigates the performance of an ASR system that uses the logarithmic, mel, or the ERB scales or a scale out of a set of similar scales when a translational VTLN is used. All scales were constructed with *nonuniform B-splines* (NURB) (Piegl and Tiller, 1997). The scales are a mapping from channel indices to filter center frequencies with an equidistant distribution on the given scale. They are illustrated in Figure 4.3 for a frequency range between 40 and 8000 Hz as dashed lines. For constructing the scales, a B-spline scheme with two fixed control points and one variable control point was used. Let the control points be defined as $P_0 = (0,0)$, $P_1 = (x_1, x_2)$, and $P_2 = (\omega, \omega)$, where ω is the Nyquist frequency and $0 \leq x_{1,2} \leq \omega$. Furthermore, let w denote a weighting coefficient in the following. The used parametric warping scheme $C(u) : [0,1] \to \mathbb{R}^2$ is then given by

$$C(u_k) = \frac{2(u_k - u_k^2)\begin{bmatrix} x_1 \\ x_2 \end{bmatrix} + u_k^2 \begin{bmatrix} \omega \\ \omega \end{bmatrix}}{(1-u_k)^2 + 2w(u_k - u_k^2) + u_k^2} = \begin{bmatrix} c_{1,k} \\ c_{2,k} \end{bmatrix}, \tag{4.13}$$

where $u_k \in [0,1], \quad k = 1, \ldots, K$. Given Equation (4.13), a number of K center frequencies $f_c(1), f_c(2), \ldots, f_c(K)$ as depicted in Figure 4.3 for given scale parameters x_1, x_2, w is computed by choosing an index dependent parameter u_k for each frequency $f_c(k)$ such that the resulting components $c_{1,k}$, $1 \leq k \leq K$, are evenly spaced within a given frequency range. Then, the corresponding components $c_{2,k}$ are used as the filter center frequencies $f_c(k)$. The warping scheme in Equation (4.13) can be used to approximate the mapping from channel indices to

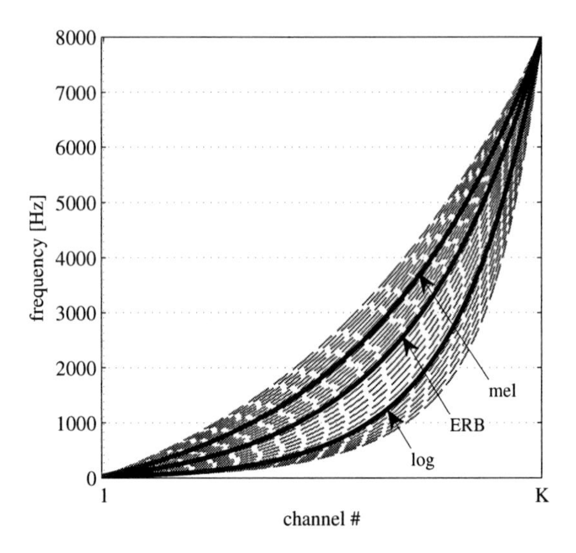

Figure 4.3.: Visualization of logarithmic, ERB, and mel scales (solid) together with alternative scales (dashed) for a frequency range of 40 to 8000 Hz.

center frequencies for a logarithmic, the mel, or the ERB scale. By using a squared error objective function, control points x_1, x_2, and the weight w for each of the three scales were computed for the experiment that is described in the following and are listed in Table 4.1. Here, a frequency range from 40 to 8000 Hz was employed. The three scales are indicated in Figure 4.3 as thick lines. Having determined approximation for these scales, a set of 32 scales was generated by interpolation of the parameters x_1, x_2, and w. The complete list of parameters for each scale in Figure 4.3 is given in Appendix C.

All scales were evaluated with phoneme recognition experiments. Here, the TIMIT corpus with the standard (matching) training-test datasets was used. Three-state triphones with up to 16 mixtures where used together with a bigram language model. Thirteen cepstral coefficients were computed that were based on the nonlinearly compressed magnitudes of the spectral values as output of different filter banks. In all experiments, 24 center frequencies between 40 and 8000 Hz were equally spaced on the scales as shown in Figure 4.3. Triangular-shaped weights and the magnitudes of gammatone filters as weights were considered. A power-law nonlinearity of 0.1 was used. The cepstral coefficients were concatenated with delta

Table 4.1.: Approximated control points and weight for the log, mel, and ERB scale when using NURB warping.

	approximated parameter		
scale	x_1	x_2	w
log	6438	34	1.6
mel	5479	1258	1.2
ERB	5891	642	1.3

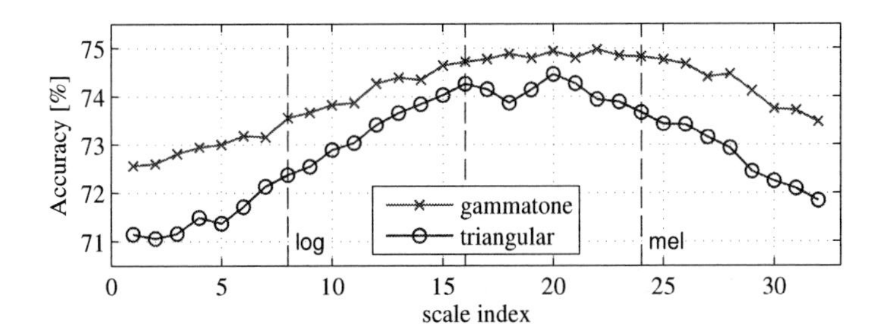

Figure 4.4.: Recognition accuracies for different frequency scales and translational VTLN on TIMIT. Triangular-shaped filters (circles) and magnitudes of gammatone filters (crosses) were used as weights. Scale index 8 refers to the log scale, scale index 16 refers to the ERB scale, and scale index 24 refers to the mel scale.

and delta-delta features to form a 39-dimensional feature vector. Furthermore, the feature vectors were mean and variance normalized (see also Section 3.5 on page 62). In contrast to the commonly used scaling VTLN (see Section 2.5.1) translational VTLN is used to compensate for the effects of different VTLs in this experiment. Speaker-adaptive training and the decoding scheme as described by Welling et al. (2002) was used and the final recognition accuracy computed as described in Section 2.4.4.

The results of this experiment are shown in Figure 4.4. Overall, it can be observed that the gammatone filter based cepstral coefficients yield higher accuracies with all scales compared to the triangular-based cepstral coefficients. In case of the triangular-based cepstral coefficients, the highest accuracy is achieved with the warping scale whose similarity lies between that of the ERB and the mel scale. The difference between the highest and lowest accuracies is about three percentage points. Using the magnitudes of gammatone filters as filter weights, this difference is about two percentage points. The highest accuracies when using gammatone filters are again obtained with scales that are similar to the ERB scale, the mel scale, or a scale in between. With respect to the accuracies depending on the filter types, these results show that both the ERB and the mel scale yield accuracies that are significantly higher than those when a logarithmic warping scale is used. This is (another) confirmation for the use of the mel or the ERB scale in conjunction with translational VTLN as described in Section 2.5.1. Schlüter et al. (2007) did not observe a better performance of cepstral coefficients based on a gammatone filter bank, but showed that the system combination of mel and gammatone filter bank based cepstral coefficients yields performance improvements of about 12 percent. However, no VTLN was used in the experiments of Schlüter et al. (2007) so that the results of the experiments described here are not fully comparable to those of that work. In the following, the experiments with invariant feature types use a gammatone filter bank and the ERB scale for locating the center frequencies of the filters.

Given the assumption that different VTLs are approximately mapped to translations along the subband-index space of the filter bank output, the following three sections describe approaches for extracting features for ASR with the use of different translation-invariant transforms.

4.4. Transformations of the Class $\mathbb{C}T$

A general class of translation-invariant transforms was originally introduced by Wagh and Kanetkar (1977) and later given the name $\mathbb{C}T$ by Burkhardt and Müller (1980). In the following, a feature extraction method for ASR is described that makes use of transforms of the class $\mathbb{C}T$. The method was published by Müller and Mertins (2010b).

The computation of transforms of the class $\mathbb{C}T$ is based on a generalization of the linear, fast *Walsh-Hadamard transform* (WHT). The WHT H_d can be defined recursively with

$$H_n = \frac{1}{\sqrt{2}} \begin{bmatrix} H_{n-1} & H_{n-1} \\ H_{n-1} & -H_{n-1} \end{bmatrix}, \qquad H_0 = 1, \tag{4.14}$$

where the normalization factor $1/\sqrt{2}$ can be omitted if appropriate for the task. By using a divide-and-conquer algorithm the WHT can be computed with a complexity of $\mathcal{O}(n \log n)$ and is then called fast WHT. Transforms of the class $\mathbb{C}T$ extract

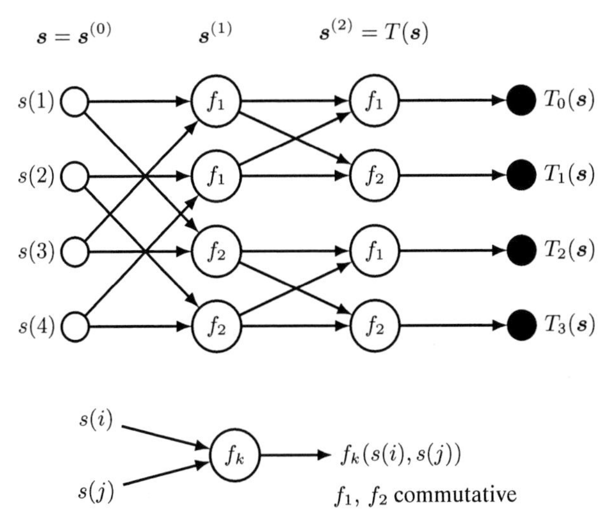

Figure 4.5.: Exemplary signal flow diagram of transformations of the class $\mathbb{C}T$ for signals with length $K = 4$.

features that are invariant under translations with cyclic boundary conditions. Given a vector $\boldsymbol{s} := (s(1), s(2), \ldots, s(K))^{\mathrm{T}}$ with $K = 2^M$, $M \in \mathbb{N}^+$ as input, members of the class $\mathbb{C}T$ are defined by the following recursive transform T with commutative operators $f_1(\cdot, \cdot)$, $f_2(\cdot, \cdot)$:

$$T(\boldsymbol{s}) := \left(T(f_1(\boldsymbol{s}_{1|2}, \boldsymbol{s}_{2|2})),\, T(f_2(\boldsymbol{s}_{1|2}, \boldsymbol{s}_{2|2}))\right), \tag{4.15}$$

where $\boldsymbol{s}_{1|2}$ and $\boldsymbol{s}_{2|2}$ denote the first and second halves of the vector $\boldsymbol{s}$, respectively,

$$\begin{aligned} \boldsymbol{s}_{1|2} &:= (s(1), s(2), \ldots, s(K/2)), \\ \boldsymbol{s}_{2|2} &:= (s(K/2+1), s(K/2+2), \ldots, s(K)). \end{aligned} \tag{4.16}$$

A binary operator $*(a, b)$ is said to be commutative if

$$a * b = b * a, \qquad \forall a, b. \tag{4.17}$$

The recursion stops with $T(s(i)) = s(i)$. Figure 4.5 shows a corresponding signal-flow diagram for $K = 4$.

The pairs of commutative operators that are examined in this work for the application in ASR were used in various pattern recognition tasks before (Wagh and

Table 4.2.: Common pairs of commutative operators for transforms of the class $\mathbb{C}T$.

operator	transform name RT	MT	QT
f_1	$a+b$	$\min(a,b)$	$a+b$
f_2	$\lvert a-b \rvert$	$\max(a,b)$	$(a-b)^2$

Kanetkar, 1977; Burkhardt and Müller, 1980). One representative of the class $\mathbb{C}T$ is the *rapid transform* (RT, Reitboeck and Brody (1969)), which has also found application in different works (for example, Wang and Shiau, 1973; Gamec and Turan, 1996; Burkhardt and Siggelkow, 2001). In comparison to the RT, it was shown by Wagh and Kanetkar (1977) that by taking the $\min(\cdot,\cdot)$ and $\max(\cdot,\cdot)$ functions as f_1 and f_2, respectively, where

$$\min(a,b) := \begin{cases} a & \text{if } a \leq b, \\ b & \text{else,} \end{cases} \qquad \max(a,b) := \begin{cases} a & \text{if } a \geq b, \\ b & \text{else,} \end{cases} \tag{4.18}$$

a higher separability compared to that of the RT can be achieved. The transform that makes use of the minimum and maximum functions as commutative operators is denoted as MT (Wagh and Kanetkar, 1977)in the following. It was shown by Burkhardt and Müller (1980) that the power spectrum of the modified WHT can be computed with a transform out of the class $\mathbb{C}T$ by choosing $f_1 := a+b$ and $f_2 := (a-b)^2$. This transform is denoted as QT. The mentioned transforms together with their according pairs of commutative operators are listed in Table 4.2. Burkhardt and Müller (1980) deduce the complete set of invariants of the class $\mathbb{C}T$ and show that the cyclic and the dyadic permutations are a subset of these invariants. Fang and Häusler (1989) describe properties of the RT, which include the additional invariance under reflection. They present a preprocessing operator $b(\cdot,\cdot,\cdot)$ for the RT that destroys the unwanted property of invariance under reflection. This operator works element-wise, with circular boundary conditions and is defined as

$$s'(i) = b(s(i), s(i+1), s(i+2)) := s(i) + |s(i+1) - s(i+2)|. \tag{4.19}$$

This particular preprocessing followed by the RT is called *modified rapid transform* (MRT), the final transform with preprocessing is then given by

$$T(\boldsymbol{s}') = T(b(\boldsymbol{s})). \tag{4.20}$$

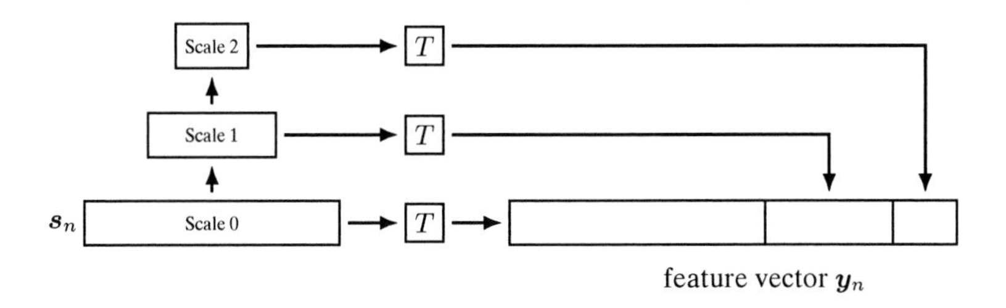

Figure 4.6.: Procedure for computing $\mathbb{C}T$-based features for ASR with multiple scales.

4.4.1. Translation-Invariant Feature Candidates for ASR

The translation-invariant features are computed on basis of the output of an auditory filter bank. A spectral value of this output is denoted as $s_n(k)$ in the following. Here, n is the frame index, $1 \leq n \leq N$, and k is the filter index with $1 \leq k \leq K$. All spectral values of a time instance n are referred to as a frame $\boldsymbol{s}_n$. In this work, a transform of the class $\mathbb{C}T$ is applied in different ways onto a frame $\boldsymbol{s}_n$. With a number of frame components K that is a power of two, such a transform can be applied directly onto a frame. Furthermore, multi-scale representations of the frames were considered as inputs to the transforms. The method of multi-scale analysis has been successfully applied to various fields of speech processing (Pinkowski, 1993; Stemmer et al., 2001; Mesgarani et al., 2004; Zhang and Zhou, 2004). Therefore, multiple scales of spectral resolution of each frame were computed. The length of each frame of scale n was half of the length of scale $n-1$. Each scale was used as input to the transform and the resulting transformations of each scale were concatenated. This procedure is illustrated in Figure 4.6 for an input of size $K = 8$. Following this procedure, the resulting number of features for an input of size $K = 2^M$ is $2^{M+1} - 1$. The concatenated features are denoted with the subscript "Scales" in the following. In addition to the transformations described above, the combination with translation-invariant VTLI features (see Section 4.2) from Mertins and Rademacher (2006); Rademacher et al. (2006) is also considered in these experiments. For the experiments, also a subset of 50 features of the "Scales"-versions of the features was determined by applying a feature selection method according to Peng et al. (2005). These feature sets are denoted with the subscript "Scales-50" in the following.

4.4.2. Experiments

On the basis of the transforms described in the previous section, different feature sets have been defined and evaluated in a number of phoneme recognition experiments. The experiments have been conducted using the TIMIT corpus with a sampling rate of 16 kHz. To avoid an unfair bias for certain phonemes, we chose not to use the "SA" sentences in training and testing similar to Lee and Hon (1989). Training and testing sets were both split into female and male utterances. This was used to simulate matching and mismatching training and testing conditions with respect to the mean VTL. Three different training and testing scenarios were defined: Training and testing on both male and female data (FM-FM), training on male and testing on female data (M-F), and training on female and testing on male data (F-M). According to Lee and Hon (1989), 48 phonetic models were trained, and the recognition results were folded to yield 39 final phoneme classes that had to be distinguished. For the acoustic modeling monophone models with three states per phoneme, eight Gaussian mixtures per state and diagonal covariance matrices were used together with bigram language statistics. MFCCs were used to obtain baseline recognition accuracies. The MFCCs were computed by using the standard HTK setup, which yields 12 coefficients and a single log-energy feature for each frame. For comparison with a VTLN technique, the method of Welling et al. (2002) was used.

We chose to use a gammatone filter bank (Patterson, 2000) with 90 filters whose center frequencies are equally-spaced on the ERB scale as basis for computing the translation-invariant features. This setup was chosen to allow for a comparison with the works from Mertins and Rademacher (2006); Rademacher et al. (2006). The frame length was 20 ms. The frame shift of the feature extraction was 10 ms. Because the transforms of the class $\mathbb{C}T$ require the length of the input data to be a power of two, the output of the filter bank was frame-wise interpolated to 128 data points. A power-function nonlinearity with an exponent of 0.1 was applied in order to resemble the nonlinear compression found in the human auditory system.

The following feature types proposed by Mertins and Rademacher (2006) and Rademacher et al. (2006) were investigated in addition to class-$\mathbb{C}T$ features: The first 20 coefficients of the DCT of the correlation term from Equation (4.11) with $d = 0$ (denoted as "ACF") have been used, as well as the first 20 coefficients of the DCT of the Equation (4.12) with $d = 4$ (denoted as "CCF"). The features belonging to the class $\mathbb{C}T$ as described in the previous section were considered together with their "Scales" and "Scales-50" versions. In order to limit the size of the resulting feature vectors, the "Scales-50" versions were used for feature-set combinations with four and five different feature types.

Table 4.3.: Phoneme recognition accuracies [%] of individual feature types.

Feature type	FM-FM	M-F	F-M
MFCC	66.6	55.0	52.4
RT	58.4	55.3	52.0
MRT	57.9	53.9	50.8
QT	53.0	48.0	46.1
MT	60.0	56.5	54.5
$\text{RT}_{\text{Scales}}$	64.3	57.4	56.7
$\text{MRT}_{\text{Scales}}$	64.3	58.9	58.4
$\text{QT}_{\text{Scales}}$	62.6	56.8	55.3
$\text{MT}_{\text{Scales}}$	64.1	58.8	58.0
$\text{RT}_{\text{Scales-50}}$	64.5	55.5	54.3
$\text{MRT}_{\text{Scales-50}}$	64.1	55.7	54.0
$\text{QT}_{\text{Scales-50}}$	62.3	53.1	52.2
$\text{MT}_{\text{Scales-50}}$	64.2	53.8	52.4
ACF	58.9	47.0	48.8
CCF	62.5	54.5	53.4

All invariant feature sets were concatenated with the logarithmized energy feature (see Section 2.2.4) together with delta and delta-delta coefficients. The resulting features were reduced to 47 features with LDA. The scatter matrices that were computed for the LDAs were based on the 48 phoneme classes that are contained in both the male and female utterances. This target dimensionality was chosen to allow for the comparison with previous works from Mertins and Rademacher.

All of the previously described feature types were tested individually for the three training-test scenarios. The resulting accuracies of these experiments are shown in Table 4.3. It can be seen that the MFCCs have the highest accuracy for the FM-FM scenario compared to the other considered feature types. The features resulting from the RT and MRT obtain similar accuracies compared to those of the MFCCs in the mismatching scenarios, but perform worse in the matching scenario. The inclusion of different scales in the feature sets leads to accuracies that are comparable to those of the MFCCs in the FM-FM scenario and already outperform the MFCCs in the mismatching scenarios M-F and F-M. Using only the 50 best features from the "Scales"-feature sets leads to accuracies that are similar to the feature sets that include all scales. However, in the mismatching scenarios the "Scales-50" versions perform worse than the "Scales" version. The correlation-based VTLI features perform similar in comparison to the $\mathbb{C}T$-based ones. In comparison, the cross-correlation features CCF perform better than the auto-correlation features

Table 4.4.: Highest phoneme recognition accuracies for feature sets with different sizes and energy amendment.

Feature type combination + energy	FM-FM	M-F	F-M
MT_{Scales}+CCF	65.7	61.1	60.5
MRT+CCF	65.4	60.6	60.5
MRT_{Scales}+CCF+ACF	65.9	61.8	61.9
MRT_{Scales}+MT_{Scales}+CCF	65.7	62.0	61.9
$MRT_{Scales\text{-}50}$+CCF+ACF+$RT_{Scales\text{-}50}$	66.0	61.3	60.6
$MRT_{Scales\text{-}50}$+CCF+ACF+$RT_{Scales\text{-}50}$+$QT_{Scales\text{-}50}$	65.9	61.8	61.2

ACF. This indicates the importance of contextual information for ASR.

As a further baseline performance, the VTLN method (Welling et al., 2002) using MFCCs as features has been tested on the three scenarios. Since this method adapts to the vocal tract length of each individual speaker, it gave the best performance in all cases. The results were as follows: FM-FM: 68.6%, M-F: 64.0%, F-M: 63.4%.

To investigate in how far the performance of the translation-invariant features can be increased through the combination of different feature types, all possible combinations of the "Scales"-versions of the features and the ACF and CCF features have been considered and combined with each other by concatenating the corresponding feature vectors. Two, three, four, and five feature types have been combined with each other for these experiments. The concatenated feature vectors were reduced with LDA to a 47 dimensional feature vector. The results for the best combinations are shown in Table 4.4.

As the results show, the combination of two well-selected feature sets leads to an accuracy that is comparable to the MFCCs in the matching case. However, in contrast to the MFCCs, feature-type combinations lead to an accuracy that was 5.6 to 7 percentage points higher in the M-F scenario and 8.1 to 9.5 percentage points higher in the F-M scenario. In particular, the results indicate that the information contained in the CCF features is quite complementary to that contained in the $\mathbb{C}T$-based features. Also the MRT and MT features seem to contain complementary information. The observation that the accuracies do not increase by considering combinations of four or five feature sets could either be explained by the fact that the "Scales-50" features in comparison to the "Scales" features have a much lower accuracy for the gender separated scenarios or by the assumption that the RT, MRT and QT do contain similar information.

In a third experiment, we supplied the previously considered, $\mathbb{C}T$-based translation-invariant feature combinations with MFCCs, as this had been necessary to boost

Table 4.5.: Highest phoneme recognition accuracies for feature type combinations of different sizes with concatenated MFCC features.

Feature type combination + MFCC	**FM-FM**	**M-F**	**F-M**
$\mathrm{MT}_{\mathrm{Scales}}$+CCF	65.6	61.6	61.5
$\mathrm{MRT}_{\mathrm{Scales}}$+CCF+ACF	66.5	62.1	62.4
$\mathrm{MRT}_{\mathrm{Scales\text{-}50}}$+CCF+ACF+$\mathrm{RT}_{\mathrm{Scales\text{-}50}}$	66.3	61.9	61.2
$\mathrm{MRT}_{\mathrm{Scales\text{-}50}}$+CCF+ACF+$\mathrm{RT}_{\mathrm{Scales\text{-}50}}$+$\mathrm{QT}_{\mathrm{Scales\text{-}50}}$	66.5	61.9	62.0

the performance with the methods described by Mertins and Rademacher (2006); Rademacher et al. (2006). The results of the experiment are shown in Table 4.5. It is notable that the MFCCs do not increase the accuracies significantly in the matching scenarios and increase only slightly in the mismatching scenarios. This means that the MFCCs do not carry much more additional discriminative information compared to the feature set combinations that solely consist of translation-invariant features.

Using the best feature set presented in the work from Rademacher et al. (2006) leads to the following accuracies: FM-FM: 65.7%, M-F: 60.8% and F-M: 59.9%. As expected, these results indicate a better performance in the gender separated scenarios than the MFCCs. However, the $\mathbb{C}T$-based translation-invariant feature sets perform even better and do not rely on the MFCCs.

4.5. Generalized Cyclic Transformations

The *generalized cyclic transformation* (GCT) is another translation-invariant transform and was proposed within the field of image analysis by Lohweg and Müller (2001); Lohweg (2003). The transform yields translation-invariant features and consists of two steps: First, the input signal is linearly transformed by a generalized characteristics matrix, which is explained in detail below. Second, a translation-invariant spectrum is computed from the transformed input signal. A method that makes use of this transform for extracting features for ASR is explained in detail next. It was published by Müller et al. (2009).

4.5.1. Definition of the Generalized Cyclic Transformation

In this section the class of generalized cyclic transformations is described following the presentation by Lohweg and Müller (2001). For this, let $\boldsymbol{s} \in \mathbb{R}^K$ be an input vector and let $\hat{\boldsymbol{s}} \in \mathbb{R}^N$ be the transformation of $\boldsymbol{s}$. These vectors are related by the transformation matrix $\boldsymbol{A}_K \in \mathbb{R}^{K \times K}$ as

$$\hat{\boldsymbol{s}} = \boldsymbol{A}_K \cdot \boldsymbol{s}. \tag{4.21}$$

The notion of *negacyclic matrices* (Davis, 1979), also known as skew circulants, is instrumental in the following sections. A negacyclic matrix $\boldsymbol{C} \in \mathbb{R}^{K \times K}$ is defined by a coefficient vector $\boldsymbol{c} := (c_0, c_1, \ldots, c_{K-1})$ as

$$\boldsymbol{C}[\boldsymbol{c}] := \begin{bmatrix} c_0 & c_1 & \cdots & \cdots & c_{K-1} \\ -c_{K-1} & c_0 & c_1 & \cdots & c_{K-2} \\ \vdots & \vdots & \ddots & \ddots & \vdots \\ & & & \ddots & \\ -c_2 & -c_3 & -c_4 & \ddots & c_1 \\ -c_1 & -c_2 & -c_3 & \cdots & c_0 \end{bmatrix}. \tag{4.22}$$

By introducing matrices $\boldsymbol{T}_{K/2}, \boldsymbol{T}_{K/4}, \ldots, 1$, with $\boldsymbol{T}_K \in \mathbb{R}^{K \times K}$, the transformation matrix $\boldsymbol{A}_K$ of the GCT can be defined recursively as

$$\boldsymbol{A}_K := \begin{bmatrix} \boldsymbol{T}_{K/2} & -\boldsymbol{T}_{K/2} \\ \boldsymbol{A}_{K/2} & \boldsymbol{A}_{K/2} \end{bmatrix}, \tag{4.23}$$

where $\boldsymbol{A}_1 := 1$. The derivation of the following result can be found in the work of Lohweg and Müller (2001). Based on the observations by Ahmed et al. (1971), the idea is to choose the matrices $\boldsymbol{T}_{K/2}, \boldsymbol{T}_{K/4}, \ldots, 1$ such, that the *absolute value spectrum* (AVS) of $\hat{\boldsymbol{s}}$, which is introduced in the next subsection, stays unchanged under cyclic translation of $\boldsymbol{s}$. It can be shown that this is fulfilled by defining

$$\boldsymbol{T}^{\boldsymbol{b}_M} := -\boldsymbol{C}[\boldsymbol{b}_M], \tag{4.24}$$

where $\boldsymbol{b}_M = (b_0, b_1, \ldots, b_{M-1})$ is a coefficient vector. In this context, the matrix $\boldsymbol{T}$ is called a *generalized cyclic matrix* (GCM). A set of $\log_2(K)$ coefficient vectors

$$\left\{ {}^{(1)}\boldsymbol{b}_{K/2}, {}^{(2)}\boldsymbol{b}_{K/4}, \ldots, {}^{\log_2(K)}\boldsymbol{b}_1 \right\} \tag{4.25}$$

defines a transformation matrix $\boldsymbol{A}_K$. The concatenation $\tilde{\boldsymbol{b}}$, $\tilde{\boldsymbol{b}} \in \mathbb{R}^K$, of these vectors is referred to as the "characteristic coefficient vector",

$$\tilde{\boldsymbol{b}} := \left[{}^{(1)}\boldsymbol{b}_{K/2}^{\mathrm{T}} \quad {}^{(2)}\boldsymbol{b}_{K/4}^{\mathrm{T}} \quad \cdots \quad {}^{\log_2(K)}\boldsymbol{b}_1^{\mathrm{T}} \right]^{\mathrm{T}}. \tag{4.26}$$

Now, by unfolding the recursion in Equation (4.23), the transformation matrix $\boldsymbol{A}_K$ can be written as

$$\boldsymbol{A}_K := \begin{bmatrix} \boldsymbol{T}^{(1)}\boldsymbol{b}_{K/2} & & & \boldsymbol{0} \\ & \boldsymbol{T}^{(2)}\boldsymbol{b}^2_{K/4} & & \\ & & \ddots & \\ \boldsymbol{0} & & & 1 \end{bmatrix} \cdot \left(\prod_{i=1}^{\log_2(K)-1} \operatorname{diag}\left(\boldsymbol{I}_{K-2^i}, \boldsymbol{H} \otimes \boldsymbol{I}_{2^{i-1}}\right) \right) \cdot \left(\boldsymbol{H} \otimes \boldsymbol{I}_{K/2}\right), \tag{4.27}$$

where diag$(\,\cdot\,,\,\cdot\,)$ defines a diagonal block matrix with two submatrices. $\boldsymbol{I}_K$ is the identity matrix of size K, and $\boldsymbol{H}$ is a Hadamard matrix of order 2,

$$\boldsymbol{H} := \begin{bmatrix} +1 & -1 \\ +1 & +1 \end{bmatrix}. \tag{4.28}$$

For example, with $\tilde{\boldsymbol{b}} = (\beta_0, \beta_1, \ldots, \beta_7)$ the transformation matrix $\boldsymbol{A}_8$ would be given by

$$\boldsymbol{A}_8 = \left[\begin{array}{cccc|cccc} -\beta_3 & -\beta_2 & -\beta_1 & -\beta_0 & +\beta_3 & +\beta_2 & +\beta_1 & +\beta_0 \\ +\beta_0 & -\beta_3 & -\beta_2 & -\beta_1 & -\beta_0 & +\beta_3 & +\beta_2 & +\beta_1 \\ +\beta_1 & +\beta_0 & -\beta_3 & -\beta_2 & -\beta_1 & -\beta_0 & +\beta_3 & +\beta_2 \\ +\beta_2 & +\beta_1 & +\beta_0 & -\beta_3 & -\beta_2 & -\beta_1 & -\beta_0 & +\beta_3 \\ \hline -\beta_5 & -\beta_4 & +\beta_5 & +\beta_4 & -\beta_5 & -\beta_4 & +\beta_5 & +\beta_4 \\ +\beta_4 & -\beta_5 & -\beta_4 & +\beta_5 & +\beta_4 & -\beta_5 & -\beta_4 & +\beta_5 \\ \hline -\beta_6 & +\beta_6 & -\beta_6 & +\beta_6 & -\beta_6 & +\beta_6 & -\beta_6 & +\beta_6 \\ \hline -\beta_7 & -\beta_7 & -\beta_7 & -\beta_7 & -\beta_7 & -\beta_7 & -\beta_7 & -\beta_7 \end{array}\right]. \tag{4.29}$$

Indicated by the dashed lines, Equation (4.29) also illustrates the recursive structure of the transformations matrices $\boldsymbol{A}_K$ that leads to a period-wise order of the rows.

It is noteworthy that the rationalized MWHT and the *square wave transformation* (SWT) (Pender and Covey, 1992) can be realized with the GCT by choosing appropriate characteristic coefficients. Besides the coefficients for the MWHT and the SWT, three additional generating rules for the coefficients were proposed by Lohweg et al. (2004). The GCT-related experiments within this work are based on these. An overview of the considered coefficients is given in Table 4.6.

Table 4.6.: Characteristic coefficient vector for different types of GCT transforms.

Transform	**Coefficient vector $\tilde{b}$**
SWT	$(\underbrace{1,1,\ldots,1}_{K-3 \text{ coeffs.}},-1,-1,1)$
MWHT	$(\underbrace{0,0,\ldots,0,-1}_{K/2 \text{ coeffs.}},\underbrace{0,0,\ldots,0,-1}_{K/2^2 \text{ coeffs.}},\ldots,-1)$
C_1	$(2^{K/2-1},2^{K/2-2},\ldots,2^0,2^{K/4-1},2^{K/4-2},\ldots,2^0,\ldots,$ $0,-1,-1,-1)$
C_2	$(\tilde{b}_1,\tilde{b}_2,\ldots,\tilde{b}_K)$ with $\tilde{b}_k=-1/K\cdot\cos(\pi\cdot(k+1/2)/K)),$ $k=1,\ldots,K-1,\ \tilde{b}_K=1$
C_3	$(r_0,r_1,\ldots,r_{K-1})\,,\ r_k\in\mathcal{N}(0,1)$

4.5.2. Translation-Invariant GCT-Features

With the given transformation $\hat{\boldsymbol{s}}$ of $\boldsymbol{s}$, translation-invariant features are computed by summarizing the coefficients that correspond to the same period in $\boldsymbol{A}_K$. Two approaches are described in this subsection. While the first method is based on the AVS, the second method uses the cyclic autocorrelation function for defining a translation-invariant extended group spectrum.

Similar to the computation of the power spectrum of the Walsh-Hadamard transformation (Ahmed and Rao, 1975), an AVS for a GCT transformed signal can be defined. It is defined as the period-wise addition of the absolute values of the transformed signal $\hat{\boldsymbol{s}}$. Formally, let

$$F_i := K\left(1-0.5^i\right),\quad i\in\mathbb{N}_0, \tag{4.30}$$

be a supplementary function that is used for ease of notation in the following. Then, $\text{AVS}(\hat{\boldsymbol{s}}):\mathbb{R}^K\to\mathbb{R}^{\log_2(K)+1}$ with $\text{AVS}(\hat{\boldsymbol{s}})=(a_0,a_1,\ldots,a_{\log_2(K)})$ is defined as

$$a_i := \begin{cases} \sum_{k=F_i}^{F_{i+1}-1}|\hat{s}_k|, & i=0,\ldots,\log_2(K)-1,\\ |\hat{s}_K|, & i=\log_2(K). \end{cases} \tag{4.31}$$

A second translation-invariant feature type can be defined on base of a cyclic autocorrelation function. Based on Ahmed et al. (1973), the cyclic autocorrelation for the GCT is defined as

$$\boldsymbol{R}_{\hat{\boldsymbol{s}}\hat{\boldsymbol{s}}}^{\boldsymbol{U}} := \begin{bmatrix} \boldsymbol{U}_0 & & & 0 \\ & \boldsymbol{U}_1 & & \\ & & \ddots & \\ 0 & & & 1 \end{bmatrix} \cdot \hat{\boldsymbol{s}}, \tag{4.32}$$

where

$$\boldsymbol{U}_i := \boldsymbol{C}\big[\big(\hat{s}_{F_i}, \hat{s}_{F_i+1}, \ldots, \hat{s}_{F_{i+1}-1}\big)\big], \qquad i = 0, 1, \ldots, \log_2(K)-1. \tag{4.33}$$

According to Lohweg and Müller (2002), the autocorrelation of a GCT transformed signal is highly redundant. However, by introducing the signum function into Equation (4.32), a translation-invariant spectrum, denoted as *extended group spectrum* (EGS), can be computed. The EGS : $\mathbb{R}^K \to \mathbb{R}^K$ of a transformed signal $\hat{\boldsymbol{s}}$ is defined as

$$\mathrm{EGS}(\hat{\boldsymbol{s}}) := \boldsymbol{R}_{\hat{\boldsymbol{s}}\hat{\boldsymbol{s}}}^{\boldsymbol{V}}, \tag{4.34}$$

where

$$\boldsymbol{V}_i := \boldsymbol{C}\big[\mathrm{sgn}\,\big(\hat{s}_{F_i}, \hat{s}_{F_i+1}, \ldots, \hat{s}_{F_{i+1}-1}\big)\big], \quad i = 0, 1, \ldots, \log_2(K)-1. \tag{4.35}$$

The drawback of the EGS function is the higher dimension of its image compared to the AVS. It is noteworthy that the AVS is contained within the EGS (Lohweg and Müller, 2002).

4.5.3. Application of GCT for feature extraction in ASR

For the application of GCTs for an ASR task, a time-frequency (TF) representation of an input speech signal has to be computed with a (quasi-)logarithmic frequency scale. This maps the spectral effects due to different VTLs approximately to translations along the frequency axis.

The way the GCT-features are computed here is inspired from Lohweg et al. (2004). For each frame the GCT is applied on all subframes according to a chosen subframe length and a subframe shift. An example for a frame of with eight components, a chosen subframe length of four and a subframe shift of two is shown in Figure 4.7. The combination of a chosen subframe length and subframe shift is called a *subframing scheme* in the following.

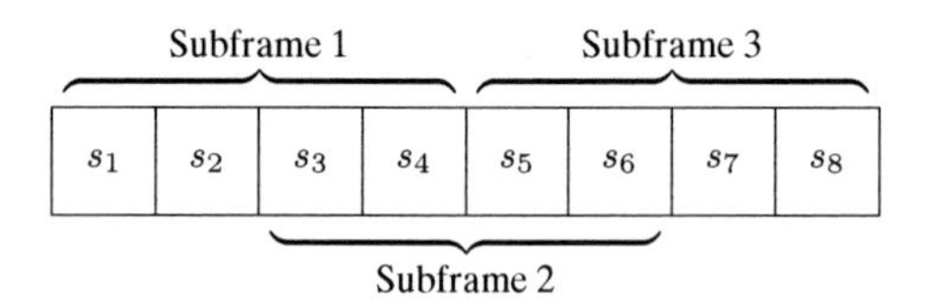

Figure 4.7.: Exemplary application of the GCT to subframes of length four and a subframe shift of two.

4.5.4. Phoneme Recognition Experiments with GCT-based Features

In order to evaluate the different schemes for choosing the characteristic coefficient vectors (see Table 4.6) and to allow for a comparison between different subframing schemes, phoneme recognition experiments on the TIMIT corpus were conducted.

Experimental Setup The "SA" sentences of the TIMIT corpus have not been used to avoid an unfair bias for certain phonemes (Lee and Hon, 1989). In order to simulate mismatching training and testing conditions with respect to the mean VTL, the training and testing data was split into male and female subsets and three scenarios were defined:

1. Training on both male and female data and testing on male and female data (FM-FM),
2. training on male data and testing on female data (M-F) and
3. training on female data and testing on male data (F-M).

For the TF analysis a gammatone filter bank (Patterson et al., 1992) was used. The number of filters was set to 64 and corresponds to the choice made by Sinha and Umesh (2002) who conducted experiments with a translational VTLN method. We used this number of filters as compromise between frequency resolution and size of feature vector. The minimum center frequency was 40 Hz, the maximum center frequency was set to 8 kHz. The final frame length was set to 20 ms and the frame shift was 10 ms. The bandwidth was chosen as 1 ERB for each filter. A power-function nonlinearity with an exponent of 0.1 was applied in order to resemble the nonlinear compression found in the human auditory system.

The recognizer was based on the hidden-Markov model toolkit (HTK) (Young et al., 2009). Monophone models with three states per phoneme, 8 Gaussian mixtures per state and diagonal covariance matrices were used together with bigram statistics. According to Lee and Hon (1989), 48 phonetic models were trained and the recognition results were folded to yield 39 final phoneme classes that had to be distinguished. All feature vectors were supplemented with the first and second order time derivatives. When the size of the feature vector was greater than 47 elements, a linear discriminant analysis was performed such that the feature vectors were reduced to a length of 47.

MFCCs were used to obtain baseline recognition accuracies. They were computed by using the standard HTK setup, which yields 12 coefficients plus the logarithmized energy for each frame. The accuracies for the three scenarios when using MFCCs in this setup are as follows: FM-FM: 66.6%, M-F: 55.0% and F-M: 52.4%. It can be seen that the accuracy of the MFCCs declines significantly when the training and testing conditions do not match with respect to the mean VTL.

Different Characteristic Coefficient Vectors and Subframing Schemes The first part of the experiments evaluated the two feature types AVS and EGS for the characteristic coefficients shown in Table 4.6. Three different subframing schemes were considered in these experiments:

1. (trivial) subframing scheme with subframe length of 64, which means that the GCT is applied on the full frame,
2. subframe length of 16 and subframe shift of 16,
3. subframe length of 16 and subframe shift of 8.

With respect to the accuracy, a general observation for the experiments is that all five coefficient-generating functions from Table 4.6 lead to very similar performances. Thus, the results of the transformation C_1 are shown in Table 4.7 as a representative.

Overall, it can be observed that the EGS feature type performs better than the AVS feature type for a given subframing scheme. Using the trivial subframing scheme, both feature types perform worse then the MFCCs. However, using a nontrivial subframing scheme increases the accuracies of both the AVS- and the EGS-based features. The accuracy of the EGS for the mismatching training-testing scenarios is slightly higher than the one of the MFCCs that was about 55% and 52% for the M-F and F-M scenarios, respectively. In comparison with the nonoverlapping subframing scheme, the subframing scheme with overlapping subframes shows slightly higher accuracies for both feature types.

Table 4.7.: Phoneme recognition accuracies (for different training/test scenarios) of C_1 transformation as representative for different subframing schemes.

Subframe length 64, subframe shift 0			
	FM-FM	**M-F**	**F-M**
AVS	52.0	49.2	46.5
EGS	57.1	52.6	50.0
Subframe length 16, subframe shift 16			
	FM-FM	**M-F**	**F-M**
AVS	56.6	50.7	49.6
EGS	62.8	55.7	54.9
Subframe length 16, subframe shift 8			
	FM-FM	**M-F**	**F-M**
AVS	59.0	51.5	51.2
EGS	64.2	56.1	56.2

Combining Translation-Invariant Feature Types Previous works showed that combinations of translation-invariant feature types increase the robustness of features in mismatching training-testing conditions (Mertins and Rademacher, 2006; Müller and Mertins, 2009b). The second part of the experiments thus investigated combinations of the GCT-based feature types. In addition, feature types based on the autocorrelation and cross correlation (Mertins and Rademacher, 2006, see also Section 4.2) of frames were considered in this experiment as well. These feature types are denoted as ACF and CCF, respectively. Again, all five coefficient types as shown in Table 4.6 were considered for the GCT-based features. Based on the results described in the previous section, the best subframing scheme was used. Therefore, a subframe length of 16 and a subframe shift of eight was chosen. For all considered coefficient types, the EGS was computed as GCT-based feature type. All possible combinations of the GCT- and correlation-based feature types with size two, three and four have been evaluated. In addition, the last part of this experiment concatenates the considered combinations with MFCCs. This was necessary in the work of Mertins and Rademacher (2006) to boost the performance of the translation-invariant feature types. Table 4.8 shows the combinations that led to the highest accuracies within the described experiments.

Generally, it can be observed that the feature type that is based on the cross correlation of two frames ("CCF") is always part of the best feature-type combinations. This indicates the importance of contextual information for the feature extraction in

Table 4.8.: Phoneme recognition accuracies of feature-type combinations.

	Feature types	FM-FM	M-F	F-M
	C_3 + CCF	65.5	59.5	59.6
	MWHT + CCF	65.5	59.4	60.0
	C_3 + ACF + CCF	65.7	60.7	60.6
	MWHT + ACF + CCF	65.8	60.9	60.5
	ACF + CCF	63.2	58.1	56.8
	C_1 + C_2 + ACF + CCF	66.0	60.6	60.7
	C_3 + SWT + ACF + CCF	66.0	60.4	60.9
+MFCC	C_3 + CCF	66.5	59.8	59.2
+MFCC	C_3 + ACF + CCF	66.6	60.8	61.2
+MFCC	C_1 + C_2 + ACF + CCF	66.6	61.1	61.6

ASR systems. Results for the best combinations of two feature types are shown at the top of Table 4.8. Compared to the accuracies that were achieved with individual GCT-based features, the accuracies in all scenarios further increased to around 65.5% in the matching and slightly less than 60% in the mismatching training-testing conditions. The introduction of a third feature type increases the accuracies in all three scenarios slightly. Again, it can be observed that the highest accuracies with combinations of three feature types were achieved by including the correlation based feature types ACF and CCF. Supplementary, the results for the combination of ACF and CCF features are shown in the middle part of Table 4.8. The fact that the combination of only ACF and CCF leads to lower accuracies in the scenarios shows that the GCT-based feature types do contain additional discriminative information. The combination of four feature types yields no further improvements.

As shown in the last part of Table 4.8 the accuracies of the scenarios do not change significantly when the translation-invariant features are concatenated with MFCC features. While the concatenation step was necessary when using only the correlation based feature types ACF and CCF as described by Mertins and Rademacher (2006) it was shown in Section 4.4 that the combination of correlation based features and another translation-invariant feature type did not benefit from additional MFCCs (see also Müller and Mertins, 2009b). The same observation is made here. The GCT based feature types in combination with ACF and CCF seem to include the discriminative information of the MFCCs.

4.6. Contextual Invariant Integration Features

In this section a third method for the computation of invariant features is described. In contrast to the first two methods describes in the sections above, however, the third approach makes use of a constructive approach for extracting invariants. This allows for the extraction of features that are invariant to specific groups of transforms and has found application in many different areas of pattern recognition (for example, Schulz-Mirbach, 1995b; Siggelkow, 2002; Temerinac et al., 2007; Reisert, 2008). The presented method was published in the works of Müller and Mertins (2009a, 2010a, 2011a).

As with the previous two method described in the following relies on a TF analysis that approximately maps the spectral scaling due to different VTLs to translations along the subband-index space. With respect to the notions introduced in Section 4.1, this translation effect can be attributed to the action of the group G of translations. With this assumption, the TF representations S of two speakers A and B of the same utterance are related as formulated in Equation (2.109). The same mathematical notations as in the previous sections are used here and are repeated in the following for convinience. Formally, let $s_n(k)$ denote the TF representation of a speech signal, where n is the time index, $1 \leq n \leq N$, and k is the subband index with $1 \leq k \leq K$. A *frame* for time index n is then given by $\boldsymbol{s}_n = (s_n(1), s_n(2), \ldots, s_n(K))^{\mathrm{T}}$. When considering a translation according to α_T in the subband-index space, some boundary conditions need to be introduced. Periodic boundary conditions, where all subband indices are understood modulo K, have been used by Müller and Mertins (2009a, 2010b) in case of CT- and GCT-based features, because they were required by the applied invariance transforms. In the following, we use repeated boundary conditions $s_n(k) = s_n(1)$ for $k < 1$ and $s_n(k) = s_n(K)$ for $k > K$, because they form a closer match to frequency warping in the analog domain.

By assuming that a finite set of translations along the subband-index space describes the occurring spectral effects due to different VTLs sufficiently[1], a finite group of translations $\widehat{G}$ with $|\widehat{G}|$ elements can be defined. For reasons of simplicity, integer translations are used in the following. Nevertheless, noninteger translations could be used if an appropriate interpolation scheme would be incorporated in the following definitions. According to the "integration approach" as described in Section 4.1, Equation (4.6) from page 75 becomes

$$\widehat{T}_f(\boldsymbol{y}) = \frac{1}{|\widehat{G}|} \sum_{g \in \widehat{G}} f(g\boldsymbol{y}). \tag{4.36}$$

[1] This is similar to the assumptions made by a typical grid-search based VTLN method, where a warping factor is chosen out of a finite set of possible warping factors.

The question of how to define the function f arises. According to Noether's theorem (Noether, 1915; Schulz-Mirbach, 1992, 1995b), in case of a finite group of translations a complete transformation

$$\widehat{T}(\boldsymbol{y}) = \left(\widehat{T}_{f_1}(\boldsymbol{y}), \widehat{T}_{f_2}(\boldsymbol{y}), \dots, \widehat{T}_{f_F}(\boldsymbol{y})\right)^{\mathrm{T}} \tag{4.37}$$

can be constructed by only considering monomials for the kernel functions f. Given the vectors $\boldsymbol{k} = (k_1, k_2, \dots, k_M)$ and $\boldsymbol{l} = (l_1, l_2, \dots, l_M)$, containing element indices and integer exponents with $\boldsymbol{k} \in \mathbb{N}^M$ and $\boldsymbol{l} \in \mathbb{N}_0^M$, respectively, a noncontextual monomial $m(n; w, \boldsymbol{k}, \boldsymbol{l})$ with M components is defined in the following as

$$m(n; w, \boldsymbol{k}, \boldsymbol{l}) := \left[\prod_{i=1}^{M} s_n^{l_i}(k_i + w)\right]^{1/\gamma(m)}, \tag{4.38}$$

where $w \in \mathbb{N}_0$ is a spectral offset parameter that is used for ease of notation in the following definitions. The value $\gamma(m)$ enotes the *order of a monomial* m:

$$\gamma(m) := \sum_{i=1}^{M} l_i. \tag{4.39}$$

The all-encompassing exponent $1/\gamma(m)$ in Equation (4.38) acts as a normalizing term with respect to the order of the monomial. Noether showed that for input signals of dimensionality K and finite groups with $|\widehat{G}|$ elements, the group averages of monomials with order less or equal $|\widehat{G}|$ form a generating system of the input space. Such a basis has at most $\binom{|\widehat{G}|+K}{K}$ elements. It has to be pointed out that this is an upper bound, and it was shown in many applications that the number of practically needed basis functions is considerably smaller (see, for example, Schulz-Mirbach, 1992, 1995b; Siggelkow, 2002). We can assume that the maximal translation that occurs as effect of VTL changes in the subband-index space is limited to a certain range W (Sinha and Umesh, 2002). A noncontextual *invariant-integration feature* (IIF) $A_m(n)$ for a frame at time n, as a group average on the basis of a monomial$m$, is defined as

$$A_m(n) := \frac{1}{2W+1} \sum_{w=-W}^{W} m(n; w, \boldsymbol{k}, \boldsymbol{l}), \tag{4.40}$$

with $W \in \mathbb{N}_0$ determining the window size. To give an explanatory example, we consider a monomial of order three with exponents $l_i = 0$ for all $i \in \{1, 2, \dots, K\} \setminus \{k_1, k_2, k_3\}$ and $l_i = 1$ for $i \in \{k_1, k_2, k_3\}$. The corresponding IIF $A_m(n)$ with

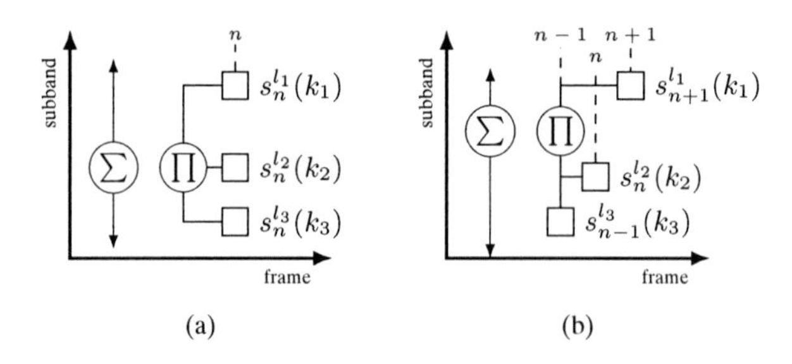

Figure 4.8.: (a) Schematic plot of a noncontextual IIF with exponents $l_1, l_2, l_3 \neq 0$, (b) schematic plot of a contextual IIF with exponents $\boldsymbol{l} = (l_1, l_2, l_3)$, $l_1, l_2, l_3 \neq 0$ and corresponding temporal offsets $\boldsymbol{m} = (m_1, m_2, m_3) = (+1, 0, -1)$.

window parameter $W = 1$ is then given by

$$A_m(n) = \frac{1}{3}\big[s_n(k_1 - 1)s_n(k_2 - 1)s_n(k_3 - 1) + s_n(k_1)s_n(k_2)s_n(k_3) \\ + s_n(k_1 + 1)s_n(k_2 + 1)s_n(k_3 + 1)\big]. \tag{4.41}$$

Figure 4.8 (a) shows a schematic plot of the computation of an IIF as defined in Equation (4.40).

A monomial for frame n following the definition in Equation (4.38) is only evaluated on the components of frame n and temporal context is not considered. Since the spectral translation between different speakers is assumed to be time independent the definition of a monomial in Equation (4.38) can be extended such that it also considers neighboring frames for its computation. The resulting feature type with contextual monomials is called *contextual IIF* here. Given a vector $\boldsymbol{m} \in \mathbb{N}_0^M$ containing temporal offsets, a *contextual monomial* $\hat{m}$ with M components is defined as

$$\hat{m}(n; w, \boldsymbol{k}, \boldsymbol{l}, \boldsymbol{m}) := \left[\prod_{i=1}^{M} s_{n+m_i}^{l_i}(k_i + w)\right]^{1/\gamma(\hat{m})}. \tag{4.42}$$

Consequently, a contextual IIF $A_{\hat{m}}(n)$ is then given by replacing m by $\hat{m}$ in Equation (4.40):

$$A_{\hat{m}}(n) := \frac{1}{2W+1} \sum_{w=-W}^{W} \hat{m}(n; w, \boldsymbol{k}, \boldsymbol{l}, \boldsymbol{m}). \tag{4.43}$$

With this definition, the noncontextual IIFs are a special case of the contextual ones. A schematic plot for the computation of the contextual IIFs according to Equation (4.43) is shown in Figure 4.8 (b). Following Noether's theorem as described above, an adequately chosen IIF set

$$\mathcal{A} := \{A_{\hat{m}_1}, A_{\hat{m}_2}, \ldots, A_{\hat{m}_F}\} \tag{4.44}$$

yields features that, on the one hand, are invariant to the translational spectral effects due to different VTLs, and, on the other hand, allow for discriminating between the individual classes.

The contextual IIFs can be applied on any TF representation that fulfills the assumption that spectral scaling is mapped to translation. As also discussed in Section 4.3 this is approximately the case when a mel or an ERB scale is used for locating the frequency centers within the TF analysis. This assumption was also made in the works of Umesh et al. (2002b); Monaghan et al. (2008). Depending on the application, a mean normalization can be applied to the features of each utterance to reduce the effect of channel sensitivity.

4.6.1. Feature Selection for Invariant-Integration Features

The set of parameters of the IIFs, consisting of the window size, element indices, exponents, and temporal offsets causes a huge number of possible combinations. Generally, with F features, T possible temporal offsets, B possible window sizes, K subbands, and a maximum order of D, the total count C of possible IIF sets is given by

$$C = \binom{B \cdot \sum_{d=1}^{D} (K \cdot T)^d}{F}. \tag{4.45}$$

Thus, depending on the choice of parameter constraints, a count of more than 10^{28} different feature sets is possible in practice. For finding a good subset of IIFs, an appropriate feature selection has to be done. As stated above, Noether's theorem gives an upper bound for the order of the monomials that is needed to construct a basis for the observation space. Hence, one constraint for the feature selection that can be defined is an assumption about the maximum number of group elements, which corresponds to the maximum number of different translations that can be observed in our setting. Although the maximum reasonable monomial order is constrained by this theorem, the experiments of this work show that already an order of up to three leads to good recognition accuracies.

Because of the high degree of freedom and the large amount of data, a requirement for the feature-selection method is a high computational efficiency. We used a feature selection method based on the so-called *feature finding neural network* (FFNN, Gramss (1991)). This method was applied successfully in the field of speech recognition (see, for example, Gramss, 1991; Kleinschmidt and Gelbart, 2002; Kleinschmidt, 2002) especially in case of small sets of training data. The FFNN approach works iteratively with a single-layer perceptron at its basis. A fast training of this linear classifier is guaranteed by its closed-form solution. Let $\boldsymbol{Y} \in \mathbb{R}^{N \times d}$,

$$\boldsymbol{Y} = \begin{bmatrix} \boldsymbol{y}_1 & \boldsymbol{y}_2 & \ldots & \boldsymbol{y}_N \end{bmatrix}^{\mathrm{T}}, \tag{4.46}$$

where N denotes the number of feature vectors and d is the number of features. Furthermore, let $\boldsymbol{B} \in \mathbb{R}^{N \times L}$,

$$\boldsymbol{B} = \begin{bmatrix} \boldsymbol{b}_1 & \boldsymbol{b}_2 & \ldots & \boldsymbol{b}_N \end{bmatrix}^{\mathrm{T}}, \tag{4.47}$$

be a "one-hot" matrix, where L denotes the number of classes. The matrix $\boldsymbol{B}$ indicates for each feature vector in $\boldsymbol{Y}$ its corresponding class label,

$$\boldsymbol{b}_i(l) = \begin{cases} 1, & \boldsymbol{y}_i \text{ does belong to class } l, \\ 0, & \boldsymbol{y}_i \text{ does not belong to class } l, \end{cases} \qquad 1 \leq i \leq N, \quad 1 \leq l \leq L. \tag{4.48}$$

Then a linear classifier $\boldsymbol{A} \in \mathbb{R}^{d \times L}$ that minimizes the error E in a minimum least-squares sense,

$$E = \| \boldsymbol{B} - \boldsymbol{Y}\boldsymbol{A} \|^2 \xrightarrow{\boldsymbol{A}} \min, \tag{4.49}$$

is given by

$$\boldsymbol{A} = \underbrace{(\boldsymbol{Y}^{\mathrm{T}}\boldsymbol{Y})^{-1}\boldsymbol{Y}^{\mathrm{T}}}_{=:\boldsymbol{Y}^{\dagger}} \boldsymbol{B} \tag{4.50}$$

$$= \boldsymbol{Y}^{\dagger}\boldsymbol{B}. \tag{4.51}$$

Here, $\boldsymbol{Y}^{\dagger}$ is also known as the *pseudo-inverse* of $\boldsymbol{Y}$. It should be emphasized that the linear classifier does not need to form complex decision boundaries, but to generalize well (Gramss, 1991). The feature-selection method can be summarized in the following four steps:

1. Start with a set of $F + 1$ features whose parameters are randomly chosen.
2. Use the linear classifier for computing the relevance of each feature.
3. Remove the feature with the least relevance.

4. If a stopping criterion (for example, the total number of iterations) is not fulfilled, add a new, randomly generated feature to the current feature set and go back to the second step, otherwise stop.

During the feature selection, the parameter set that leads to the highest mean relevance is memorized and returned at the end of the selection process. The linear classifier performs a frame-wise phone classification. The relevance of a feature i is computed on basis of the computation of the mean squared error of the linear classifier. The relevance of feature i is defined as the difference between the error when using all features and the error without feature i. With Equations (4.49) and (4.51) it follows that the error E can be computed with

$$E = (\boldsymbol{B} - \boldsymbol{Y}\boldsymbol{Y}^{\dagger}\boldsymbol{B})^{\mathrm{T}}(\boldsymbol{B} - \boldsymbol{Y}\boldsymbol{Y}^{\dagger}\boldsymbol{B}) \tag{4.52}$$

$$= N - \underbrace{\boldsymbol{B}^{\mathrm{T}}\boldsymbol{Y}}_{:=\boldsymbol{C}}(\underbrace{\boldsymbol{Y}^{\mathrm{T}}\boldsymbol{Y}}_{:=\boldsymbol{D}})^{-1}\underbrace{\boldsymbol{Y}^{\mathrm{T}}\boldsymbol{B}}_{=\boldsymbol{C}^{\mathrm{T}}} \tag{4.53}$$

$$= N - \boldsymbol{C}(\boldsymbol{D})^{-1}\boldsymbol{C}^{\mathrm{T}} \tag{4.54}$$

Let ${}^{(k)}\boldsymbol{C}$ denote the matrix $\boldsymbol{C}$ without column k and let ${}^{(k,k)}\boldsymbol{D}$ denote the matrix $\boldsymbol{D}$ without column k and without row k. Having computed the matrices $\boldsymbol{C}$ and $\boldsymbol{D}$, the error when feature i is discarded can efficiently be computed with

$$E = N - {}^{(k)}\boldsymbol{C}\left({}^{(k,k)}\boldsymbol{D}\right)^{-1} {}^{(k)}\boldsymbol{C}^{\mathrm{T}}. \tag{4.55}$$

It follows that this approach yields a ranking of the features according to their relevance. For determining the relevance, again, different setups are possible. For example, it can be determined by using a matching (with respect to the mean VTL) training-test scenario. Another possibility would be to compute the relevance for each feature as the mean relevance for different mismatching training-test scenarios in which, for example, male utterances are used for training and female ones are used for testing, and vice versa. The experimental part evaluates both ways of relevance computation.

4.6.2. Experiments: Data and Setup

A series of experiments has been conducted with the contextual IIFs. In the following, the results are compared with those for MFCCs and gammatone filter bank based cepstral coefficients (GTCC). First, IIFs with monomials of order one are considered. Then it is investigated whether the feature selection method used to find IIFs can also be applied to select MFCCs from a time window. Experiments with

IIFs of higher order are considered afterwards. In a second part of the experiments, speaker adaptation based on VTLN and MLLR is included into the ASR system. The generality of the selected feature sets is investigated in the last part of the experiments.

The experiments were conducted on the TIMIT and the TIDigits corpora. Details about the corpora are given in Appendix B. Similar to the feature-selection process described in Section 4.6.1, different training-test scenarios were defined for both corpora in order to simulate matching and mismatching training-test conditions. The matching scenario in TIMIT refers to the standard training and test sets. Two mismatching scenarios were defined for TIMIT. The first used only the female utterances from the training set for training and the male utterances from the test set for testing. Similarly, the second setup used only the male utterances from the training and the female utterances from the test set. Accordingly, the mismatching scenarios are denoted as "F-M" or "M-F" in the following. In the M-F setting, the amount of test data is reduced to one third of the complete test set, so that the obtained accuracies are less statistically significant than for the complete test set used in the matching scenario. However, the number of utterances is still more than twice of the often used core test set (Halberstadt, 1998), so that statistical significance of M-F results is not a serious issue that could lead to a misinterpretation of the properties of different feature sets.

For the TIDIGITS corpus, two scenarios were defined. The first used only the corresponding utterances by adults for training and test. The second scenario used the utterances by adults for training and those by children for testing. These two scenarios are denoted in the following as "A-A" and "A-C", respectively.

While these experiments explicitly look at the VTL as source of variability, other speaker dependencies also affect the results. For example, the corpora contain different dialects and speaking rates, so that good results on these corpora also indicate robustness to such variabilities. Of course, in general, one would want robustness to even more types of variation, which cannot be studied with these corpora.

The TF analysis methods differed for the considered feature types: In case of the MFCCs, the standard HTK setup was used that consists of a 26-channel mel filterbank (Davis and Mermelstein, 1980; Young et al., 2009). A gammatone filterbank implemented with an FFT approach (Ellis, 2009) was used in case of the GTCCs and IIFs. The minimum center frequency of the filters was set to 40 Hz, and the maximum center frequency was set to 8 kHz. With these constraints, the center frequencies were evenly spaced on the ERB scale. In case of GTCCs, 26 channels were used. In order to have sufficiently many spectral values for the computation of the IIFs, the number of channels was chosen as 110 in this case. Certainly, the number of channels could be smaller if the parameter for the window size would be noninteger and if an appropriate interpolation scheme would be used. However, minimizing the number of channels is not the scope of this article. The frame length was set to 20 ms and a frame shift of 10 ms was chosen. A power-function nonlinearity with an exponent of 0.1 was applied on the spectral values in order to resemble the nonlinear compression found in the human auditory system.

The recognizer was based on the *Hidden-Markov model toolkit* (HTK) (Young et al., 2009) in all experiments. Phone recognition experiments were conducted on the TIMIT corpus. State-clustered, cross-word triphone models with diagonal covariance modeling were used. All models had three emitting states with a left-to-right topology. Additionally, a long silence and a short silence model were included. A bigram language model was used that was computed on base of the TIMIT training data. According to Lee and Hon (1989) the phone-recognition results were folded to yield 39 final classes. The number of Gaussian mixtures was optimized for both feature types. While for the cepstral coefficient based feature types a mixture of 16 Gaussians per state was chosen, it turned out to be beneficial to use mixtures of eight Gaussians when using IIFs as features. Word recognition experiments were conducted with TIDigits. Here, whole-word left-to-right HMMs without skips over the states were trained. The number of states was chosen according to the average length of the individual words and varied between 9 and 15 states. A mixture of up to eight Gaussian distributions was used for all states.

For baseline accuracies, MFCCs and GTCCs with 12 coefficients plus log-energy together with first and second order derivatives were computed. Cepstral mean subtraction and variance normalization were performed. As a standard VTLN method, the approach described by Welling et al. (2002) was used. It estimates maximum-likelihood warping factors based on whole utterances with a grid-search approach. The considered warping factors α_S were chosen from the set $\{0.88, 0.9, \ldots, 1.12\}$ in case of MFCCs. For the GTCCs, the warping parameters α_T were chosen empirically from $\{-1.5, -1.25, \ldots, 1.5\}$. During training, individual warping factors were first estimated for all speakers. Then these warping factors were used to train speaker-independent models. The decoding of the test data used a two-pass strategy: First, a hypothesis was computed based on the unwarped features. Then, the hypothesis was aligned to a set of warped features, and the warping factor with the highest confidence was used for the final decoding. Speaker-adaptive training and speaker adaptation with MLLR for the tests were also considered during the experiments. When MLLR was used, a regression class tree with eight terminal nodes was employed.

4.6.3. Experiments: Feature Selection

The number of features F and the maximum order of the used monomials were varied within the individual experiments. For the feature selection, the parameters were constrained as follows: The temporal offsets m were allowed to be within an interval of ± 3 frames, which corresponds to a maximum temporal context of 80 ms. The maximum window-size parameter W was limited to 80 subbands. This number has been chosen so high to be on the safe side, and in fact, the features that were selected had values for W of up to 65. According to Equation (4.45) the total count of possible feature sets for $F = 30$ and a maximum order of 1 is of magnitude 10^{110}. Because of technical limitations, only every tenth utterance was used for the feature selection. Therefore, about 370 male and female utterances with about 110.000 frames were considered. Of course, a larger amount of data for the feature selection could lead to better generalization capabilities, but the found feature sets already show good robustness, even if an entirely different corpus is used for testing. The last part of the experiments described here investigates the generalization capabilities of the TIMIT-based feature sets by using these features sets for word-recognition experiments on the TIDigits corpus.

The result of a feature-selection process according to the described method depends on its initialization and the randomly chosen features during its runtime. Thus, for each experiment, several repetitions of the feature-selection process were made. A number of ten repetitions has been experimentally determined as appropriate number of repetitions, and an increase beyond ten did not significantly improve the results. No further heuristics for the initialization of the parameters have been used. The overall process can be described as follows:

1. Compute the TF representation of the training data as described in Section 4.6.2.

2. Perform feature selection as described in Section 4.6.1 ten times.

3. Decide for the feature set with the highest mean relevance.

With this choice of parameters, up to 15.000 different feature sets can be examined during the feature selection. The evolution of the smallest RMS error and the corresponding PER are shown for an exemplary feature selection in Figure 4.9. For this example, it can be observed that the mean RMS error correlates well with the PER for the first 400 iterations. During the following 1000 iterations, the PER increases and decreases while the best RMS error is decreasing. This behavior in the vicinity of the optimum could be expected, as the results obtained with a linear classifier can predict the performance of an HMM-based recognizer only to a limited extent. Generally, it was observed that after 1500 iterations of the described

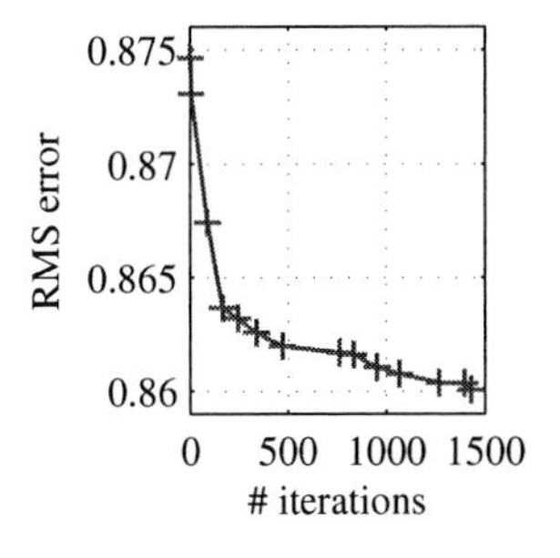

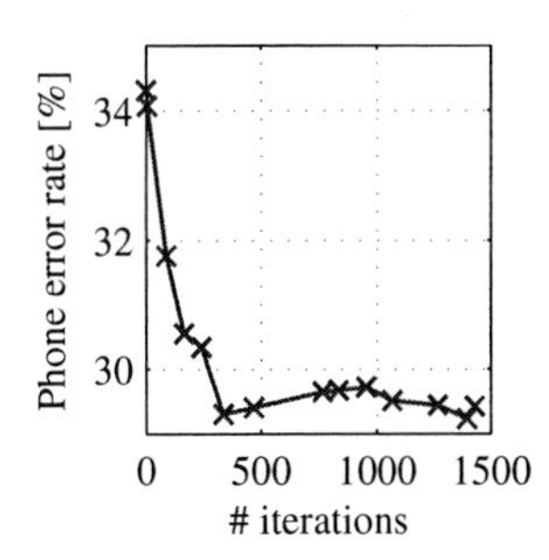

Figure 4.9.: Exemplary feature selection: (a) evolution of the smallest mean RMS error, and (b) corresponding phone error rates.

feature selection, the IIF set with the smallest mean RMS error yielded a significant improvement in accuracy compared to the randomly initialized IIF set. The IIFs that were obtained after the feature selection usually have small as well as large integration windows and involve the whole range of allowed parameter ranges. Considering the relevances of the individual features as described in Section 4.6.1, we observed that IIFs with large as well as with small integration windows were considered as highly relevant by the feature selection. Two exemplary contextual IIFs $A_{m_1}(n)$ and $A_{m_2}(n)$ that were selected during the experiments are shown in the following. They use monomials of order one, which corresponds to computing the mean spectral value for a certain frequency range,

$$\begin{aligned} A_{m_1}(n) &= \frac{1}{21} \sum_{w=-10}^{10} s_{n+1}(22 + w), \\ A_{m_2}(n) &= \frac{1}{47} \sum_{w=-23}^{23} s_{n-1}(60 + w). \end{aligned} \tag{4.56}$$

Here, the integration range for $A_{m_1}(n)$ is from about 150 Hz to about 480 Hz and for $A_{m_2}(n)$ from about 600 Hz to about 3200 Hz.

4.6.4. Experiments: Invariant Integration Features of Order One

Three IIF feature sets of order one and with sizes 10, 20, and 30 were selected in a matching scenario as described above. For the decoding, the log-energy was appended together with first and second order derivatives. Then, a *linear discriminant*

Table 4.9.: Accuracies [%] of phone recognition experiments on TIMIT for MFCCs, GTCCs, and for IIFs of order one (with $F = 10, 20, 30$). The feature vector dimensions are shown in brackets.

Features \ Scenario	match	mismatch	
	FM-FM	**F-M**	**M-F**
MFCC (39)	72.2	53.7	54.7
MFCC+VTLN (39)	73.3	67.7	70.0
GTCC (39)	72.5	55.4	54.0
GTCC+VTLN (39)	73.9	66.8	68.7
10 IIF (33)	73.4	61.6	62.9
20 IIF (55)	74.8	60.3	61.4
30 IIF (55)	75.3	60.2	61.1

analysis (LDA) was used to project the feature vectors down to 55 dimensions. Here, the phone segments were considered as individual classes (Haeb-Umbach and Ney, 1992). The dimensionality-reduction step was omitted for the feature set that consists of 10 IIFs. Finally, a *maximum likelihood linear transformation* (MLLT) (Saon et al., 2000) was applied to allow for diagonal covariance modeling. In the following, the resulting accuracies will be compared with those for MFCCs and GTCCs with and without VTLN in a phone recognition task on TIMIT. Further results on cepstral coefficient based feature types with speaker adaptation will be reported in later Section 4.6.7. Experiments with contextually enhanced MFCCs are described in the next section. To study the robustness to VTL mismatches, the usual matching case as well as the two mismatching scenarios M-F and F-M were considered for the comparison. Table 4.9 summarizes the results. The first two rows list the results for the MFCCs. By comparing the results for the MFCC matching scenario with those for the corresponding mismatching ones, it can be seen that the performance of standard MFCCs differs by about 17 percentage points. The enhancement when using MFCCs with VTLN is larger for the mismatching scenarios than for the corresponding matching scenario. The next two rows show the same behavior for GTCCs.

Analyzing the results for the IIFs, it can be seen that all three feature sets outperform the MFCCs and GTCCs without VTLN in all scenarios. Furthermore, for the matching case an increase of accuracy is observable with an increasing number of IIFs. Interestingly, the accuracies of the IIFs for the mismatching scenarios is highest for the smallest feature set consisting of 10 IIFs and is lowest with the largest feature set. An explanation may be that the larger IIF sets are better adapted to the corpus they were selected on and, therefore, do not generalize as well as the smaller

IIF set. Interestingly, all IIF feature sets yield accuracies that are comparable (10 IIFs) or even higher (20 IIFs, 30 IIFs) than the ones of the cepstral coefficient based feature types with VTLN in the matching scenario, which represents the normal mode of operation of ASR systems.

4.6.5. Experiments: Contextually Selected MFCCs

Besides approximations of first- and second order time derivatives of the feature components, another common approach for considering temporal context information is to concatenate feature vectors followed by an LDA. LDA-based combinations of MFCC and GTCC feature vectors for an 80 ms context were also evaluated. Because of their similar results, only the ones for the MFCCs are shown in the upper part of Table 4.10. The comparison with Table 4.9 shows that the accuracies for the LDA extension are higher when no speaker-adaptation is used. However, when MLLR or VTLN+MLLR are applied, the features with the time-derivative extensions yield in most cases slightly higher accuracies with our setup. The properties of the LDA-based approach have, for example, been discussed by Schlüter et al. (2006), where it was pointed out that the performance of the LDA method often drops when features are highly correlated, too many frames are concatenated, or the training set is too small. To further investigate contextually enhanced MFCCs, we will describe another approach for including contextual information now that leads to significant improvements under matching as well as under mismatching conditions.

Contextual IIFs of order one are sums of weighted spectral values where the weighting coefficients have a rectangular shape. In case of the IIFs the rectangular shape originates from the idea of integrating over the group of all possible translations. This procedure is similar to the computation of MFCCs in which the spectral values at the output of a mel filterbank are weighted with triangular shaped coefficients and integrated. For a further comparison between IIFs and MFCCs we selected 30 MFCCs from an 80 ms context using the same feature selection algorithm as for the IIFs. Similar to the IIFs, log-energy and first- and second order time derivatives were appended to the feature vectors and finally reduced with an LDA to 55 dimensions. The lower part of Table 4.10 shows the results of these experiments. The recognition rates for selected GTCCs are not listed, because they were slightly inferior to selected MFCCs.

The comparison of these results with the accuracies of the MFCCs as listed in Table 4.9 shows that the contextually selected MFCCs yield much higher accuracies than the standard MFCCs for most of the setups. It is noteworthy that the accuracies of the contextual MFCCs increase especially under mismatching conditions.

Table 4.10.: Accuracies [%] of phone recognition experiments on TIMIT for LDA-reduced concatenated MFCCs ($MFCC_{LDA}$), and for contextually selected MFCCs ($MFCC_{select}$). The feature vector dimensions are shown in brackets.

Scenario / Features	match	mismatch	
	FM-FM	F-M	M-F
$MFCC_{LDA}$ (55)	73.8	57.0	57.3
$MFCC_{LDA}$+MLLR (55)	74.2	60.4	61.2
$MFCC_{LDA}$+VTLN (55)	74.8	68.6	69.6
$MFCC_{LDA}$+VTLN+MLLR (55)	75.3	69.3	70.8
$MFCC_{select}$ (55)	74.7	60.3	61.5
$MFCC_{select}$+MLLR (55)	75.2	63.2	64.4
$MFCC_{select}$+VTLN (55)	76.1	70.9	71.2
$MFCC_{select}$+VTLN+MLLR (55)	76.4	71.2	72.1

Our conclusions of these findings are that a good feature selection of individual components within a contextual temporal window may be beneficial compared to a (linear) combination of components as it is done, for example, by an LDA. The pure selection of MFCCs from a context window has to the best of our knowledge not been proposed before.

4.6.6. Experiments: Invariant Integration Features of Higher Order and the Feature-Selection Scenario

In the next experiment, the maximum allowed order of the monomials was constrained to two and three, respectively. Furthermore, each feature selection was performed on the matching scenario, as well as on the mismatching scenarios M-F and F-M. The chosen size of the feature sets was set to 30. For comparison purposes, an IIF set of order one was also selected on the mismatching scenarios. Table 4.11 shows the results of the experiments.

It can be seen that the accuracies in the matching scenario degrade when IIFs of higher order are included, whereas the accuracies in the corresponding mismatching scenarios do increase. This effect is most noticeable when the IIFs are selected on the basis of mismatching scenarios. For example, the IIF set whose results are shown in the last row of Table 4.11 yields similar accuracies as the MFCCs when combined with VTLN. Thus, the accuracy for the normal matching cases can be traded off to increase the robustness to larger mismatches between training and testing.

Table 4.11.: Accuracies [%] of phone recognition experiments on TIMIT for 30 IIFs of higher order. "FS scenario" denotes the scenario that was used for the relevance computation during the feature selection (matching or mismatching), "order" denotes the maximum monomial order of the IIFs.

FS scenario	Order	match FM-FM	mismatch F-M	mismatch M-F
matching	1	75.3	60.2	61.1
	≤ 2	74.4	62.4	61.4
	≤ 3	74.2	62.2	61.7
mismatching	1	74.2	62.3	62.2
	≤ 2	73.3	63.3	64.9
	≤ 3	73.5	64.3	64.1

4.6.7. Experiments: Invariant Integration Features Combined with VTLN and/or MLLR

Adaptation and normalization methods are commonly part of state-of-the art ASR systems nowadays. In the following, we investigate whether the superior properties of the IIFs are still observable when VTLN and/or MLLR are also used within the ASR system. A block-diagonal structure with three blocks was set as constraint for all MLLR-transform estimations, because it turned out to be beneficial for all considered feature types. While in the experiments in Sections 4.6.4 and 4.6.6 an LDA was applied on the IIFs together with their derivatives, the experiments in this part had a different sequence of feature-processing steps: For the IIF sets of size 30, an LDA with a final dimensionality of 20 was applied. No dimensionality reduction was performed with the IIF sets of size 10 and 20. Then, the log-energy and first and second order derivatives were appended, and a 3-block-constrained MLLT was computed to decorrelate the features. Speaker-adaptive training was performed with *constrained MLLR* (CMLLR), while a combination of CMLLR and MLLR was applied in the decoding stage. When VTLN was used with IIFs, which are using a 110-channel gammatone filterbank, the considered warping parameters α_T were $-8, -7, \ldots, 8$. In case of MFCCs and GTCCs, the warping parameters were chosen as described in Section 4.6.2. Table 4.12 shows the results of the experiments.

Table 4.12.: Accuracies [%] of phone recognition experiments on TIMIT for MFCCs, GTCCs, and for IIFs of order one ($F = 10, 20, 30$) when adaptation methods are used. The feature vector dimensions are shown in brackets.

Scenario / Features	match	mismatch	
	FM-FM	**F-M**	**M-F**
MFCC+MLLR (39)	75.2	65.3	66.9
MFCC+VTLN (39)	73.3	67.7	70.0
MFCC+VTLN+MLLR (39)	75.4	69.6	71.8
GTCC+MLLR (39)	74.9	66.2	66.3
GTCC+VTLN (39)	73.9	66.8	68.7
GTCC+VTLN+MLLR (39)	76.3	70.4	72.4
10 IIF+MLLR (33)	74.9	68.0	68.6
20 IIF+MLLR (63)	75.4	67.8	69.4
30 IIF+MLLR (63)	76.2	68.1	69.4
10 IIF+VTLN (33)	75.4	69.7	70.5
20 IIF+VTLN (63)	76.2	70.1	70.1
30 IIF+VTLN (63)	77.2	71.4	70.9
10 IIF+VTLN+MLLR (33)	76.0	71.2	72.5
20 IIF+VTLN+MLLR (63)	77.1	72.4	72.3
30 IIF+VTLN+MLLR (63)	77.4	73.4	72.4

As expected, the cepstral coefficient-based systems that use both MLLR and VTLN yield higher accuracies in all scenarios than the cepstral coefficient-based systems that use only one of the adaptation methods or none of them. Generally, it can be observed that IIF-based ASR systems do benefit from the use of MLLR and/or VTLN. Compared to the accuracies from Table 4.9, it can be seen that the additional use of MLLR and VTLN increases the accuracy of the MFCC- and GTCC-based systems in the matching scenario by about 3 and 4 percentage points, respectively. It has been shown by Uebel and Woodland (1999) that the increases in accuracy due to VTLN and MLLR (in case of MFCCs) are approximately additive. This, however, was not observed by Uebel and Woodland (1999) when only CMLLR is used in combination with VTLN. This observation was explained in the work of Pitz and Ney (2005), which analytically showed that VTLN can be viewed as a special case of constrained MLLR (CMLLR) adaptation. The ASR system that is used for the experiments here makes use of CMLLR during training and a combination of CMLLR and MLLR during the decoding stage. This explains the higher accuracies for the cases in which both VTLN and MLLR were used in comparison to the cases in which only VTLN or MLLR were used. Combining

Table 4.13.: Accuracies [%] of word recognition experiments on TIDIGITS for MFCCs, GTCCs, and for IIFs of order one ($F = 10, 20, 30$). The feature vector dimensions are shown in brackets.

Features \ Scenario	match A-A	mismatch A-C
MFCC (39)	99.52	96.02
MFCC+VTLN (39)	99.59	97.25
GTCC (39)	99.65	97.67
GTCC+VTLN (39)	99.66	98.94
10 IIF (33)	99.59	97.40
20 IIF (55)	99.64	97.95
30 IIF (55)	99.68	97.89
10 IIF+VTLN (33)	99.65	99.22
20 IIF+VTLN (55)	99.73	99.38
30 IIF+VTLN (55)	99.76	99.30

MLLR and VTLN with IIFs yields increases in accuracy for the matching scenario between 2.1 and 2.6 percentage points. Even though the increase is not as large as in case of MFCCs, the IIF set of size 30 in combination with MLLR and VTLN yields the highest accuracies in the experiments. Also, in comparison to the accuracies of the contextually selected MFCCs as listed in Table 4.10 the IIF set with 30 features yields the highest accuracies in most of the setups.

4.6.8. Experiments: Experiments on TIDIGITS

Since the features were selected on base of an individual corpus (in our case TIMIT), the question arises, how good the IIF sets perform on another corpus. Therefore, the last part of the experiments considered the "generalization capabilities" of the features. The same IIF sets that were selected on the TIMIT corpus were used for the feature extraction for the word-recognition task on TIDIGITS. The results of theses experiments are shown in Table 4.13.

The first four lines of Table 4.13 show the results obtained by the cepstral coefficient-based ASR systems without adaptation and with VTLN, respectively. From line five on, the accuracies of the IIF-based systems without any adaptation and with VTLN are shown. For both scenarios, it can be seen that the IIF-based systems without VTLN yield accuracies that are equally high (10 IIF, A-A) or higher than for the MFCC-based systems. It is particularly remarkable that the IIF-based system without adaptation outperforms the MFCC-based one with VTLN even in the A-C setting, where the recognizer is trained on adult speech and tested with children speech. Due to the very low number of errors in the A-A case, the statistical significance of the results may be in question for this scenario. However, in the A-C setting, where the accuracy is generally lower, it could be improved by up to 0.7 percentage points when using IIFs instead of MFCCs with VTLN. In absolute terms, this means that the number of correctly recognized digits could be increased by the IIFs by up to 90 digits compared to MFCCs, which can be seen as a significant increase of accuracy. The comparison of IIFs with GTCCs without VTLN shows comparable accuracies. However, when VTLN is used, the accuracy of the IIFs in the A-C scenario still is about 0.4 percentage points higher than the GTCC+VTLN case, which corresponds to 30 more correctly recognized digits.

4.7. Feature Comparison and System Combination

The experimental setups that were applied during the development of the three invariant feature extraction methods as described above were different for each extraction method. Thus, a comparison by means of recognition performance of these feature types with the results presented up to here is not possible.

In the following, uniform acoustic and language models, as well as the same training and test conditions were used to evaluate all three invariant feature types under comparable conditions. In the next subsection, the features are evaluated for the cases with only individual feature types in a single ASR system. These are based on standard as well as on invariant feature types. In the subsequent subsection ROVER-combined ASR systems are evaluated. Because noise robustness of the invariant feature types is part of an individual chapter, the results of the invariant feature types on the Aurora-2 task are not presented here, but in the next chapter.

4.7.1. Single-Feature Performances

TIMIT As done for the evaluation of the standard feature types in Chapter 3, the experiment that is described in the following evaluated the performance of the invariant feature types for different training-test conditions on TIMIT: In case of matched conditions, the training and test data both contained utterances from female and male speakers. In case of mismatching training-test conditions, the training data contained data only from male speakers, and the test data contained utterances only from female speakers. Thus, a mismatch in the average VTL between training and test data was simulated. Furthermore, results are shown with and without the use of MLLR. Also here, the same ASR training and test setup as for the standard feature types was used so that the results are comparable with each other. The performance of the features with various target dimensions of the LDA was also evaluated with this experiment.

Each feature type used the parameters that lead to the best recognition performance during the feature-specific experiments as described in Sections 4.4, 4.5, and 4.6. In detail, the $\mathbb{C}T$-based features were computed by using the MRT operators together with the output of a 90-channel gammatone filterbank. Furthermore, the multi-scale approach as described in Section 4.4.1 was used. Because it lead to the highest accuracies in those experiments, a subset of the VTLI-features was concatenated to the $\mathbb{C}T$-based features. In case of the GCT features, a 64-channel gammatone filter bank was used. A subframing scheme with a subframe length of 16 components and a subframe shift of 8 components was used. Two transformation matrices with characteristic coefficients according to the schemes C_1 and C_2 (see Table 4.6) were used. The extended group spectrum was computed for each transformation. As done for the $\mathbb{C}T$-based features, a subset of the VTLI features was also concatenated to the final feature vector. For the extraction of the IIFs, a gammatone filter bank with 110 channels was used. The same 30 features that lead to the highest accuracy under matching training-test conditions as described in the experiments of Section 4.7.1 were used. All feature types appended delta and delta-delta features and a dimensionality reduction and decorrelation step with LDA and MLLT was applied. The results for the experiments on TIMIT are shown in Table 4.14.

If no MLLR is used and matching training-test conditions are considered, it can be seen that the largest accuracy is achieved by the IIFs with a target dimensionality of 55 and that the IIFs perform better than the CT and GCT features for all target dimensions. Furthermore, it can be observed that the CT features perform worst in comparison to GCT features and IIFs. In case of mismatching training-test conditions, the GCT features perform best in all cases. A reason for the worse performance of the IIFs in this case might be that a matching scenario was used for the feature selection. It was shown in Section 4.6.6 that IIFs, whose parameters were selected in a mismatching scenario, yield much better accuracies in mismatching training-test scenarios, but with the trade-off that their performance in matching training-test conditions decreases.

In addition, an advanced processing scheme was evaluated for the case, in which MLLR is used. This scheme also lead to superior performances in the experiments that were described in Section 4.6.7: It was observed in preliminary experiments that the introduction of block-matrix constraints for the estimation of MLLR transforms has a significant impact on the performance of the ASR system. The advanced processing scheme first reduced the dimension of the feature vectors down to 20 dimensions. Then, the 20-dimensional feature vectors were supplied by the log-energy

Table 4.14.: Accuracies [%] for invariant feature types on the TIMIT corpus with and without MLLR. Accuracies are shown for matching (mismatching) training-test scenarios and various target dimensions.

target dimension	Feature type	without MLLR	with MLLR
	CT	71.8 (64.6)	72.6 (65.7)
40	GCT	73.1 (65.3)	74.2 (67.1)
	IIF	74.7 (61.8)	75.7 (66.3)
	CT	72.4 (63.9)	72.8 (64.9)
50	GCT	73.8 (64.8)	74.5 (65.7)
	IIF	75.1 (61.0)	75.8 (64.9)
	CT	72.5 (64.1)	72.5 (63.9)
55	GCT	73.6 (64.6)	74.0 (66.1)
	IIF	75.3 (61.1)	75.8 (64.0)
	CT	72.3 (63.6)	72.8 (64.5)
60	GCT	73.8 (64.8)	74.1 (65.7)
	IIF	75.2 (60.9)	75.6 (63.4)

feature, as well as by the delta and delta-delta features. The subsequent MLLT and MLLR transforms were estimated with block-matrix constraints, which reduces the number of transform parameters that need to be estimated. Table 4.15 shows the recognition results for CT features, GCT features, and IIFs with this procedure.

It can be seen that the IIFs lead to the highest accuracies under both training-test conditions. The GCT features perform better than the CT features. Compared to the accuracies for the mismatching scenario as listed in Table 4.14, all three feature types benefit from the advanced processing scheme. While the introduction of the block-matrix constraints have a large impact on the accuracy on the TIMIT task, the benefits of this advanced processing scheme might vanish on larger tasks. In that case, the larger amount of available training data might allow for a more robust estimation of full transformation matrices or might even allow for the use of full covariance matrices for the acoustic modeling.

OLLO The presented invariant feature types were also evaluated on the OLLO task. The same experimental setup for OLLO that was used for the standard features in Section 3.7 was applied here. Hence, the results shown in Table 4.16 can be compared with those presented in that section. The experiments were done with

Table 4.15.: Accuracies [%] for invariant feature types with a processing scheme that has block-matrix constraints for the MLLT and MLLR transforms. The dimensionality of the final feature vectors is shown in braces.

	scenario	
Feature type	**matching**	**mismatching**
CT (63)	72.6	66.0
GCT (63)	74.1	66.9
IIF (63)	76.2	69.4

Table 4.16.: Accuracies [%] for invariant feature types on the OLLO corpus with LDA-reduction to 40 and to 55 dimensions.

target dimension	**Feature type**	**speaking rate**		
		slow	**normal**	**fast**
40	CT	57.6	64.8	53.4
	GCT	61.2	67.7	57.2
	IIF	61.8	70.9	53.1
55	CT	58.8	66.3	55.0
	GCT	**62.6**	70.3	**58.6**
	IIF	61.3	**72.1**	55.7

two different target dimensionalities for the LDA: On the one hand, a dimension of 40 (comparable to the one of the standard feature types) was used. On the other hand, a target dimension of 55 was used, which was empirically obtained in preliminary experiments. The acoustic models were trained only on utterances with a normal speaking rate.

In case of 40-dimensional feature vectors it can be observed that the IIFs achieve the highest accuracy under matching training-test conditions ("normal" speaking rate). In comparison to the best performing standard feature type PLP (see Table 3.3 on page 66), the IIFs achieve an even higher recognition rate. The second highest accuracies were obtained with GCT features. It can be observed for all three feature types that the decrease in accuracy in case of mismatching training-test conditions is larger for the utterances of the group "fast speaking rate" than for utterances of the group "slow speaking rate". As can be seen in the lower part of Table 4.16, the accuracies of each feature type can be increased by choosing a target dimension of 55. This holds for all considered training-test conditions with the exception of IIFs and a "slow" speaking rate in the test data. By comparing the highest accuracies achieved with the standard features for each condition with the corresponding highest accuracies from the invariant feature types, the results show that the invariant feature types achieve accuracies that are between 2.5 and 5 percentage points higher.

4.7.2. System Combinations

As described in Section 2.6, advanced ASR systems often combine several recognizers with each other in order to improve the overall recognition performance. Besides cross-adaptation (Gales et al., 2007), which uses the output of one system as the input of another system, another common method is to take hypotheses of multiple systems and combine them. This could be done, for example, with the *recognizer output voting error reduction* (ROVER, see also Section 2.6) method or *confusion network combination* (CNC) (Evermann and Woodland, 2000). Compared to the accuracies of individual recognition systems, the combining approaches have shown the ability for large performance improvements (Stüker et al., 2006; Schlüter et al., 2007). In this section, the outputs of the individual systems are combined with ROVER. Standard as well as invariant features as described in the previous sections are considered.

Data and Modeling Phoneme recognition experiments have been conducted on the TIMIT corpus with a sampling rate of 16 kHz. As in the previous sections two training-test scenarios have been defined: The first includes female and male utterances in the training and test set. The second scenario simulates a mismatch in the mean VTL between training and test data. Therefore, the training set of the second scenario includes only male utterances from the original training set, while the test set includes only female utterances from the complete test set. Following the standard approach (Lee and Hon, 1989), an initial set of 48 phonemes was used for the training of the acoustic models. This set was folded to 39 phonemes for testing purposes.

In all systems triphone HMMs were trained with a left-to-right three-state topology with no skip states. The output distributions were modeled with diagonal covariance matrices. Decision-tree clustering was applied for state tying and a bigram language model was used. In contrast to the TIMIT experiments in Section 4.7.1, each system also applied VTLN. For determining the warping factors, the systems based on invariant feature types used cepstral coefficients that were based on the TF representation used by the individual feature types. In case of the MFCC- and

Table 4.17.: Baseline accuracies of individual systems. The dimensionality of the feature vectors of each system is shown in brackets.

	Accuracy for scenario [%]	
Front-end type	**matching**	**mismatching**
MFCC (39)	76.0	69.7
PLP (39)	76.6	70.0
GCT (55)	74.9	69.6
VTLI (55)	75.0	66.8
CT (55)	73.0	69.4
IIF (60)	**77.5**	**72.6**

PLP-based system, scaling was used for the warping of the frequency axis, while the invariant-feature based systems used a translational VTLN. *Speaker-adaptive training* (SAT) with CMLLR and speaker-adaptation with a combination of CMLLR and MLLR during testing were employed.

Two systems with standard feature types were considered for the experiments. For the first system, MFCCs with 12 coefficients were used. The second system computed 12 PLP coefficients. All systems used a frame length of 20 ms and a frame shift of 10 ms and appended log-energy together with first and second order derivatives to the feature vectors.

The settings for the individual front-ends with invariant feature types were adapted to the settings as presented in the works of Müller and Mertins (2011a); Mertins and Rademacher (2006); Müller et al. (2009); Müller and Mertins (2010b). The parameters that yielded the highest accuracies within the individual works were taken. In case of VTLI-, GCT- and CT-based features, an LDA was used to reduce the dimension of the feature vectors to 55. In case of IIFs, 30 features of order one were selected. The final dimensionality of the IIF vectors after applying LDA was 60.

Baseline Error Rates The accuracies of the individual systems for both scenarios are shown in Table 4.17. The upper part of the table shows the results of the systems that use the standard MFCC- and PLP-based front-ends. It can be seen that the PLP-based system performs slightly better than the MFCC-based system in both scenarios. The lower part of Table 4.17 shows the accuracies of the systems whose front-ends are based on invariant feature types. The highest accuracies, which are also higher than the ones of the PLPs, are achieved with the IIFs. In contrast, the three other invariant feature types yield accuracies that are lower than the accuracies of the noninvariant MFCC- and PLP-based systems. For the mismatching scenario, it would be expected that the accuracies of the invariant-feature based systems are higher than the ones of the systems based on standard features. This is only the case for IIFs here. A reason for this shortfall may be the fact, that the back-ends were not individually optimized to the feature types. Another reason is that VTLN and MLLR do a good job especially with the noninvariant features.

System Combinations The outputs of the individual systems have been combined with the ROVER approach: After the alignment of the 1-best hypotheses into a single phoneme transition network, a subsequent module processes the network and selects the word with the best score (Fiscus, 1997). Within this work, the ROVER implementation of the NIST *Scoring Toolkit* (SCTK) (Nat, 2010) was used.

All possible combinations of the individual systems as described above were considered for this part of the experiments. Table 4.18 shows a selection of the results. This table contains the combinations of size two, three, four, five, and six with highest accuracy, as well as the best combinations of different sizes when only invariant feature types are used. Generally, it can be observed that an output combination of the considered ASR systems leads to performance improvements. The combination of systems based on noninvariant features with systems based on invariant features yields higher performance improvements than combinations of only invariant-feature based systems. It can be observed for the matching scenario that the accuracy increases with an increasing number of systems combined with each other. For the mismatching scenario, the combination of five systems is slightly better than that of six systems. Compared to the accuracies of the baseline IIF system, the error rate is reduced by about 2.2 percentage points in the matching and 2.1 percentage points in the mismatching case. This means a relative error rate reduction of 11% for the matching and 6% for the mismatching scenario.

Table 4.18.: Accuracies of system combinations with different sizes. The systems combined with each other are marked with •. The accuracies of the individual systems for matched training-test conditions are shown in braces.

MFCC	**PLP**	**GCT**	**VTLI**	**CT**	**IIF**	Accuracy [%] for scenario	
(76.0)	(76.6)	(74.9)	(75.0)	(73.0)	(77.5)	**matching**	**mismatching**
	•				•	78.0	72.9
		•			•	77.5	72.7
	•		•		•	79.3	73.8
			•	•	•	77.9	73.2
•	•	•			•	79.4	74.4
		•	•	•	•	78.5	73.8
•	•	•	•		•	79.6	**74.7**
•	•	•	•	•	•	**79.7**	74.2

5

Noise Robustness of the Invariant Feature Types

One of the major issues in the field of ASR research nowadays is the noise robust feature extraction for ASR systems (Feng et al., 2012). In the following, the noise robustness of the invariant features that were presented in the last chapter, is evaluated. An excerpt of methods for enhancing the noise robustness of ASR systems was introduced in Section 2.2.6. The described methods are feature-enhancement methods. With a first experiment in Section 5.1, no further feature-enhancement step is done so that the performance of the features can be compared with that of the nonenhanced standard features as given in Section 3.7. Section 5.2 evaluates the enhancement methods "mean and variance normalization", RASTA, and the power-normalization steps in combination with the invariant integration features. Section 5.3 describes a method that makes use of a sophisticated auditory model known as "auditory image model".

5.1. Baseline Accuracies for Noisy Conditions

As explained for the standard feature types in Section 3.7, the noise robustness of the different features was evaluated on the AURORA-2 task and, for convenience, the

Table 5.1.: Baseline word recognition accuracies [%] for the experiments on the AURORA 2 corpus for (a) clean-speech training and (b) multi-style training. Excerpt from Table 3.2, repeated for convenience here.

	SNR [dB]						
Feature type	∞	20	15	10	5	0	-5
MFCC	98.6	96.8	93.0	78.1	51.2	26.3	12.2
PNCC	98.6	97.7	95.7	90.1	75.7	49.1	21.9

(a)

	SNR [dB]						
Feature type	∞	20	15	10	5	0	-5
MFCC	98.4	97.9	96.8	93.9	85.3	65.1	31.1
PNCC	98.0	97.7	97.0	95.1	88.3	72.1	40.2

(b)

performances of MFCC and PNCC features on this task are repeated in Table 5.1. To allow for a comparison, the same acoustic and language modeling that was used for the standard feature types was also emplyed for the experiments with the invariant feature types. Whole-word models with four Gaussians and diagonal covariance matrices were used. For the CT- and GCT-based feature extraction, the parameters that lead to the highest accuracies for the experiments as described in Sections 4.4 and 4.5 were applied. The parameters for 30 IIFs were selected on the training set with the procedure described in Section 4.6.1. The constraints were chosen such that monomials of order one with a maximum temporal context of ± 3 frames were selected. The experiments with the CT- and GCT-based features showed that the correlation-based VTLI features from the work of Mertins and Rademacher (2005) increase the overall performance of the invariant features. Thus, experiments were also conducted where the CT-, GCT-, and integration-based features were concatenated with a subset of the VTLI features as explained in Section 4.4.2. For another baseline, only the VTLI features without and with concatenated MFCCs were evaluated on the AURORA task. All features were concatenated with their corresponding delta and delta-delta features followed by mean and variance normalization. In case of the invariant feature types, an LDA with target dimensionality of 55 and a subsequent MLLT were used. The results of the experiments are listed in Table 5.2.

At first, the results of the CT, GCT, and IIF features in the top three rows of Table 5.2

Table 5.2.: Accuracies [%] of invariant feature types on the AURORA 2 task for (a) clean-speech training and (b) multi-style training.

Feature type	**SNR [dB]** ∞	20	15	10	5	0	−5
CT	98.2	95.7	91.8	83.1	65.1	37.7	16.7
GCT	98.4	95.9	92.3	83.0	66.0	39.2	16.0
IIF	99.2	97.8	95.4	89.5	72.8	41.0	14.0
CT_{VTLI}	98.8	97.3	94.9	88.8	74.2	48.5	21.3
GCT_{VTLI}	99.1	97.7	95.1	88.1	72.8	46.2	20.6
IIF_{VTLI}	99.2	97.7	95.2	89.0	74.7	48.0	19.1
VTLI	98.8	96.7	93.3	85.1	68.5	40.8	15.4
$\text{VTLI}_{\text{MFCC}}$	98.8	97.1	94.2	86.7	71.7	46.3	19.3

(a)

Feature type	**SNR [dB]** ∞	20	15	10	5	0	−5
CT	97.2	95.8	94.0	90.1	80.4	57.0	25.4
GCT	97.4	96.2	94.6	89.9	79.2	55.5	24.8
IIF	98.7	98.3	97.4	95.6	89.2	68.1	31.0
CT_{VTLI}	98.3	97.5	96.5	94.1	87.6	68.0	31.8
GCT_{VTLI}	98.8	98.0	97.0	94.2	87.2	67.4	32.9
IIF_{VTLI}	98.9	98.2	97.4	95.4	89.4	72.2	36.6
VTLI	97.9	97.9	97.0	94.6	87.4	68.1	32.5
$\text{VTLI}_{\text{MFCC}}$	98.0	98.2	97.2	94.8	88.2	69.3	34.2

(b)

are compared. For clean-speech training it can be seen that the integration-based IIFs achieve the highest accuracies under the considered noise conditions with exception of an SNR of −5 dB. In case of multi-style training, the IIFs achieve the highest accuracies for all SNRs. In comparison to the MFCCs, the IIFs perform better or similar under all conditions. In comparison to PNCC features, however, the integration features perform similar (down to 15 dB) or worse (10 dB and below) in case of clean-speech training. With multi-style training, PNCC and IIF features perform similar for SNRs down to 5 dB and the IIFs perform worse for SNRs of 0 and −5 dB. The rows four to six of Table 5.2 show the accuracies for invariant features that are supplied by VTLI-features. This is indicated by the subscript "VTLI". It can be observed that the performance of the CT and GCT features increases under all noise conditions. In case of the integration features this increase is only observable for SNRs of 0 and −5 dB. Overall, the results show that the VTLI-supplemented invariant features perform very similar. Compared to the PNCCs, it can be seen that the performance of the supplemented invariant features is generally slightly worse under clean-speech training conditions and comparable under multi-style training conditions.

In another experiment the noise robustness of the VTLI features was evaluated. It was described by Mertins and Rademacher (2005) that the best performance for the VTLI-features can be observed when they are supplemented by MFCC features. Because of this, experiments were conducted that evaluate the VTLI features without and with concatenated MFCC features. The last two rows of Table 5.2 show the results of these experiments. The VTLI features with concatenated MFCCs are denoted by the subscript "MFCC". By comparing the accuracies of the VTLI features with the ones of CT, GCT, and IIF features for clean-speech training, it can be seen that the VTLI features perform better than the CT- and GCT-based features for all SNRs with exception of −5 dB. This does also hold for the multi-style training accuracies with the exception that the VTLI features perform best with an SNR of −5 dB. Supplying the VTLI features with MFCCs improves the accuracies of the VTLI features for both training styles. In comparison, however, the VTLI-supplied IIFs perform equal or better for all SNRs except −5 dB.

5.2. Feature-Space Methods for Enhanced Noise Robustness

Method Comparison The last section compared the performances of the different feature types with mean normalization as the only enhancement-step for increasing the noise robustness. In this section, the feature-enhancement method "mean normalization" and "variance normalization", as well as RASTA processing, power-normalization, and power-bias subtraction, as described in Section 2.2.6, are evaluated and compared with each other. The experiments concentrate on unsupervised feature-enhancement methods, and, thus, SPLICE is not considered in the following experiments. In case of no feature enhancement, the invariant feature types CT, GCT, and IIF follow the same extraction scheme: After a time-frequency analysis, a nonlinear compression of the magnitudes of the spectral values was done, followed by an extraction of invariant features. Then, delta- and delta-delta features were appended. Before passing the feature vectors to the back-end, an LDA and MLLT was applied to reduce the dimensionality to 55 and to decorrelate the feature vectors. With respect to this scheme, the CT-, GCT-, and IIF-extraction differ in the stage, where the nonlinear transform is applied. Because the invariant-integration features showed the highest accuracies under different noise conditions with mean normalization only (see Table 5.2), the IIFs are used in the experiments of this section as a representative invariant feature type. The same IIF parameters as selected in the previous section are used here.

The experiments were conducted on the AURORA task with the same acoustic modeling and the same language modeling as used for the standard feature types in Section 3.7. If used, the feature-enhancement methods were applied in the following way:

- Mean normalization (MN) and variance normalization (VN) were done after the computation of the the CT-, GCT-, or IIF-based features.
- Power normalization (PN) was applied directly on the power spectrum of the speech signal.
- The relative spectral (RASTA) filter was applied after the TF analysis with the filter bank. The impact of different choices for the pole p of the RASTA filter (see Equation (2.42) on page 31), as well as the impact of different non-linearities were also evaluated in combination with the IIFs. Three different poles, $p \in \{0.94, 0.96, 0.98\}$, were considered. This choice was motivated from the work of Hermansky and Morgan (1994). Instead of using a log-arithm operator, a power-function nonlinearity was used in that work. As described in more detail in Section 2.2.6, an optimal choice of this nonlinearity may depend on the noise conditions and is a trade-off between robustness against additive and convolutional noise. Furthermore, preliminary experi-ments showed that the use of the logarithm operator in combination with the gammatone filter bank, which is used for the IIF extraction, lead to worse accuracies than the ones obtained with a power-function nonlinearity. Fig-ure 5.1 shows the resulting accuracies when the pole and the nonlinearity is changed. Here, clean-speech training was used and the average accuracy for all noise conditions in the test data is presented.

  It can be seen that the choice of the pole p has only slight effects on the average accuracy of the ASR system. In contrast to that, however, the chosen exponent for the power-function has a significant impact on the recognition rate of the system. For the considered set of exponents, it can be seen that a choice of 0.18 or 0.22 (depending on the pole) yield the highest accuracies with nearly 70 percent in average. In the following experiments, an exponent of 0.22 and a pole with $p = 0.96$ was used.
- The power-bias spectral subtraction (PBS) method was also used after the TF analysis with the filter bank. The smoothing parameters, that are described in the work of Kim and Stern (2009), were empirically optimized in preliminary experiments. The enhancement of the integration features with the power-bias

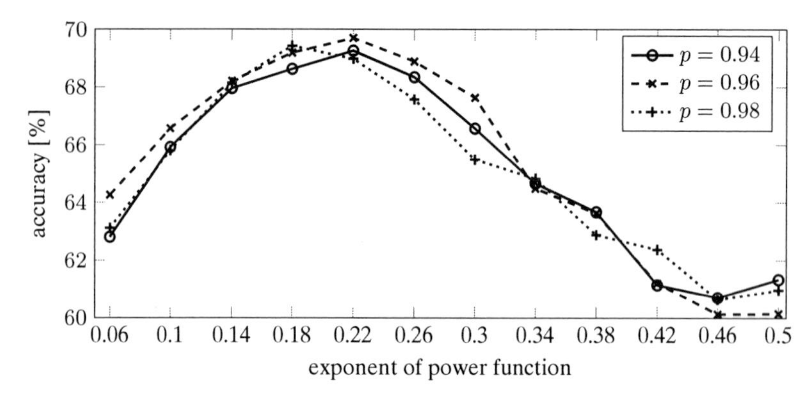

Figure 5.1.: Accuracies for IIFs with RASTA pre-processed filter bank outputs. For the RASTA filter (see Equation (2.42) on page 31) different poles p were considered, $p \in \{0.94, 0.96, 0.98\}$.

subtraction approach was also presented in the work of Müller and Mertins (2011b), where the TIMIT corpus was used as training and test dataset.

Table 5.3 shows the accuracies for invariant-integration features on the AURORA task for clean-speech training. The top row of Table 5.3 shows the accuracies that result from not employing any feature-enhancement methods. The next two rows list the accuracies for the cases in which only mean normalization (MN), or mean and variance normalization (VN) are used. When using MN only, the performance under all noise conditions increases and gives accuracies that are among the highest within this experiment. Compared to these the combination of mean and variance normalization, however, leads to accuracies that are lower under all noise conditions. Furthermore, the use of VN in combination with all other considered feature-enhancement methods also shows no performance improvements. This is a surprising result, because the use of variance normalization in combination with, for example, MFCC features is common for ASR under noisy conditions. A reason for this poor performance might be that no voice-activity detection is implemented within the used ASR framework, so that the normalization stage also considers silence segments. The use of RASTA filtering after the computation of a TF representation also does not lead to accuracies that are competitive with the ones that are obtained with MN only. A general remark can be made here: Due to the large parameter space of the IIFs (see Section 4.6), a selection of IIF parameters on a preprocessed TF representation might take possible artifacts, introduced by the feature-enhancement method, into account and might lead to increased accura-

Table 5.3.: Accuracies on AURORA with clean-speech training for invariant-integration features (IIF) with different kinds of pre-processing applied. The highest accuracy for each condition is indicated as bold text.

	SNR [dB]						
Enhancement method	∞	20	15	10	5	0	-5
-	98.6	96.8	93.3	77.9	46.6	20.1	10.7
MN	99.2	97.8	95.4	89.5	72.8	41.0	14.0
MN+VN	95.2	92.9	89.3	80.7	62.6	32.6	12.4
RASTA+MN	97.6	94.8	90.9	78.8	53.0	22.2	9.3
RASTA+MN+VN	97.2	88.0	78.4	60.1	34.7	15.4	9.4
PN+MN	98.8	97.4	95.1	88.4	69.5	37.8	14.5
PN+MN+VN	98.7	92.9	89.0	79.9	60.9	31.5	11.5
PBS+MN	**99.1**	97.6	95.4	89.6	74.1	43.6	15.9
PBS+MN+VN	98.7	88.7	77.4	56.6	62.0	16.2	10.1
PN+PBS+MN	98.8	**97.8**	**96.0**	**91.0**	**77.7**	**50.4**	**22.1**
PN+PBS+MN+VN	98.3	94.4	89.8	79.2	59.2	33.2	15.7

cies. The bottom six rows of Table 5.3 evaluate the power-normalization (PN) and power-bias subtraction (PBS) approaches. The combination of PN, PBS, and MN gives the highest accuracies under noisy conditions within these experiments. While the use of PN or PBS plus MN as only enhancement method leads to accuracies that are comparable to those when only MN is used, the combination of all three feature enhancement methods shows large improvement especially for SNRs of 10 to -5 dB. This result supports the work of Kim and Stern (2010a) and confirms the subtle selection of these feature-enhancement steps for increasing the noise robustness.

5.3. Auditory Image Model for Increased Noise Robustness

Different types of auditory filter banks were described in Section 2.2.1. As it could be observed with the experiments in Section 3.8, simply replacing one filter bank by another filter bank, possibly a more sophisticated one, does not necessarily lead to improvements in recognition accuracy. However, it was shown by Walters (2011) that the use of a more detailed model of the human auditory system can improve the robustness of an ASR system under noisy conditions. In the following, a method is shown that combines such a model with an invariance transform in order to increase the overall robustness of an ASR system under noisy conditions. It was published by Müller and Mertins (2012c).

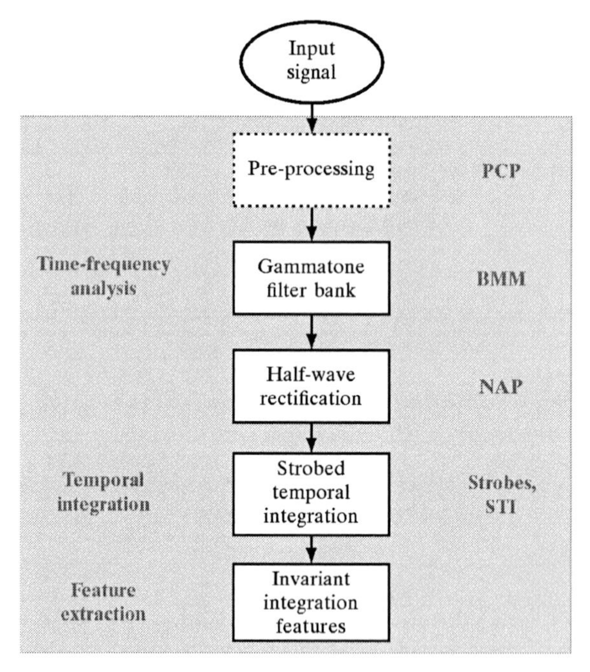

Figure 5.2.: Overview of the modular structure of the auditory image model (AIM, gray shaded area) Patterson et al. (1992), which consists of the pre-cochlear processing (PCP, not used in this work) module, basilar membrane motion (BMM) module, neural activity pattern (NAP) module, and the strobed temporal integration (STI) module, which makes use of a preceding strobe detection method.

5.3.1. The Auditory Image Model

The use of auditory representations that differ from "traditional" spectrograms are an active field of research within the field of ASR (Feng et al., 2012). The *auditory image model* (AIM) is such a computational model that simulates human auditory processing. It aims to convert the speech signal into the perception that a human initially has before any semantic meaning is associated with the signal. The AIM is divided into several modules that have either physical or psychoacoustic analogies. In the following, a brief overview of these modules and the alternative methods within the modules is given. An illustration of the core structure of the AIM is shown within the gray shaded region in Figure 5.2.

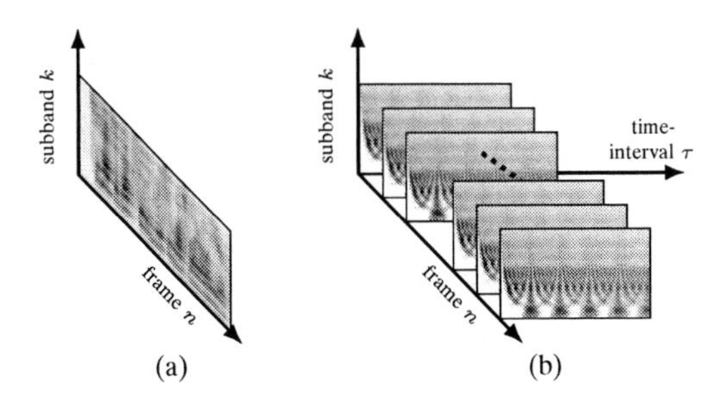

Figure 5.3.: (a) Visualization of frame-wise time-frequency representation, (b) Visualization of stabilized auditory images of a speech signal.

Pre-cochlear processing (PCP): The simulation of the transfer function from the sound field to the oval window of the cochlea is carried out in the PCP module of the AIM. In this work, however, we did not make use of this module.

Basilar membrane motion (BMM): The spectral analysis performed in the cochlea is simulated in the BMM module. The gammatone (Patterson et al., 1987) as well as the static and dynamic-compressive gammachirp (Irino and Patterson, 2006) filter bank could be used within this module. All filter banks locate the frequency centers of their filters equally spaced along the (quasi-logarithmic) ERB scale (Patterson et al., 1992). For reasons of computational efficiency the gammatone filter bank was used in the experiments that are described in the following.

Neural activity pattern (NAP): The neural transduction from the BMM to the auditory nerve is simulated in the NAP module. The unipolar response of the hair cells and (optionally) dynamic-range compression are modeled by half-wave rectification and logarithmic or square root compression. The loss of phase lock can also be simulated with a channel-wise low-pass filter here.

Temporal integration: For ASR tasks, features are commonly computed with a rate of 100 Hz. Therefore, segments ("frames") with a length of about 20-25 ms and an offset of about 10 ms are defined. For a given *time-frequency* (TF) representation, a spectral profile as basis for a subsequent feature extraction in ASR systems is generally generated by low-pass filtering and downsampling the output of the preceding filter bank. An exemplary sequence of spectral profiles is illustrated in Figure 5.3 (a).

Findings of perceptual research suggest that fine structure of periodic sounds is preserved in later stages of the auditory pathway, which cannot be explained with a running temporal integration process as described above, but rather by using a two-stage process that is referred to as *strobed temporal integration* (STI) (Patterson et al., 1992): First, peaks in the NAP have to be identified as strobe points for each channel. Ideally, these peaks correspond to the onsets of glottal pulses. Then, the strobe points are used to initiate a temporal integration process, which adds weighted segments of the NAP into the corresponding channel of a buffer. The segments start at the time of the strobe-point occurrence and have a default length of 35 ms; The values of this buffer represent the *stabilized auditory image* (SAI). The SAI values decay continuously with a default half life of 30 ms. This time span is referred to as "SAI decay time" in the following. The SAI is a two-dimensional, time-continuous perceptual representation, where one dimension is indexed by the subband number, and the other one corresponds to time intervals relative to the strobe times. Because the strobe points are chosen synchronously to glottal pulses, the resulting SAI is GPR aligned. For a given speech signal, a series of SAIs is obtained by taking SAI samples with a given frame rate. This generates a three-dimensional representation and is illustrated in Figure 5.3 (b). A more detailed description and a comparison of different strobe finding algorithms for the AIM can be found in the works of Walters (2011); Bleeck et al. (2004). In this work, we used the "constrained threshold" ("sf2003" in Bleeck et al. (2004)) algorithm for strobe finding and the "ti2003" algorithm (Bleeck et al., 2004) for performing temporal integration.

The effects of different VTLs within the SAI space have been described by Patterson et al. (2007), where it was shown that the scale-time space of the SAI is scale covariant. This means that a change of the VTL leads to a shift along the subband axis, as well as to a scaling along the time-interval axis. The scaling is caused by the different lengths of the impulse responses of the filters.

The AIM has proved to yield beneficial auditory representations for various speech processing tasks: Monaghan et al. (2008) used the AIM to extract low-dimensional features for speech recognition that proved to be more robust to VTL changes than MFCC features on a synthetic speech corpus. The experimental part of this work also considers this kind of features for comparison. In the work of Lyon et al. (2010), sparse codes from the SAI were computed and used for sound retrieval and ranking. There exist different publicly available implementations of the AIM. Here, we used the Matlab version (Bleeck et al., 2004). In the next section we describe two alternative approaches for the feature extraction, which will be analyzed in the experimental part of this thesis.

5.3.2. AIM-based Features for ASR: Cepstral Analysis and Invariant Integration

Cepstral coefficients are used in many state-of-the-art ASR systems due to their good performance and their efficient computation that involves only a few parameters. However, with respect to speaker independence, the cepstral analysis with an auditory filter bank has the disadvantage that different VTLs lead to translations along the subband axis, while the *discrete cosine transform* (DCT) is not translation invariant. Thus, the same phoneme uttered from two vocal tracts with different lengths do not yield the same point in the MFCC space. Generally speaking, a feature extraction method should only extract information that is necessary for separating the individual classes of interest and, at the same time, be invariant to the effects of other variabilities. Using the AIM for feature extraction for noise-robust ASR can further be motivated by the observations made in Walters (2011), were SAI-based features showed a larger noise robustness than MFCCs (while performing worse under clean-speech conditions).

As described in Section 4.6, invariant integration is a constructive approach for computing separable features that are invariant to a designated group of transformations. IIFs were presented in that section as a feature extraction method for ASR that makes use of this integration approach. The key concept of the IIFs is their invariance to translations along the subband axis. Motivated by the observation that the SAI space is scale covariant, we propose a new definition for invariant-integration features based on the SAI space in this work. Therefore, one has to integrate over the induced transformation due to different VTLs in this space. In the following, we first give a short formal introduction to IIFs based on a standard mel or gammatone filter bank. With the then introduced terms, it is described afterwards how SAI-based IIFs can be computed.

We follow the notation as introduced in Section 4.6: Let $s_n(k)$ denote the TF representation of a speech signal, where n is the time index, $1 \leq n \leq N$, and k is the subband index with $1 \leq k \leq K$. We define the vectors $\boldsymbol{k} = (k_1, k_2, \dots, k_M)$ and $\boldsymbol{l} = (l_1, l_2, \dots, l_M)$, containing element indices and integer exponents with $\boldsymbol{k} \in \mathbb{N}^M$ and $\boldsymbol{l} \in \mathbb{N}_0^M$, respectively. Furthermore, let $\boldsymbol{m} \in \mathbb{Z}^M$ be a vector containing temporal offsets. Now, we define a contextual monomial $\hat{m}$ with M components as

$$\hat{m}(n; w, \boldsymbol{k}, \boldsymbol{l}, \boldsymbol{m}) := \left[\prod_{i=1}^{M} s_{n+m_i}^{l_i}(k_i + w) \right]^{1/\gamma(\boldsymbol{l})}, \tag{5.1}$$

where $\gamma(\boldsymbol{l}) := \sum_{i=1}^{M} l_i$ is a normalizing term and is referred to as the "order of the monomial". Also, $w \in \mathbb{N}_0$ is a spectral offset parameter that is used for ease of notation in the following definitions. Now, a single IIF is defined as

$$A_{\hat{m}}(n) := \frac{1}{2W+1} \sum_{w=-W}^{W} \hat{m}(n; w, \boldsymbol{k}, \boldsymbol{l}, \boldsymbol{m}), \tag{5.2}$$

with W determining the window size. An adequately chosen IIF set of size F,

$$\mathcal{A} := \{A_{\hat{m}_1}, A_{\hat{m}_2}, \ldots, A_{\hat{m}_F}\}, \tag{5.3}$$

yields features that, on the one hand, are invariant to the translations, approximating spectral effects due to VTL changes, and, on the other hand, allow for discriminating between the individual classes. For determining the parameters of a set of IIFs an iterative feature selection method is used that is based on a linear classifier as described in Section 4.6.

To apply the invariant integration approach within the SAI space, one has to integrate over the induced transformation due to different VTLs within the SAI space. That is, the scaling effect between the subbands with different VTLs has to be considered. This relation can be described with the product of time interval and center frequency of the individual subbands being constant (Irino and Patterson, 2002): Let $\tilde{v}_k(n, \tau)$ denote the SAI value at time instance n, with k being the subband index, and τ being the time interval. Furthermore, let $\boldsymbol{c} = (c_1, c_2, \ldots, c_K)$ denote the center frequencies of the filters. Now, for a given subband index $i \in \mathbb{N}$ and a cycle number $p \in \mathbb{R}^+$,

$$\tau_i(p) := \frac{p}{c_i} \tag{5.4}$$

defines the time interval for each subband that corresponds to the same cycle for all impulse responses. Now, the SAIs of the same utterance from two speakers A and B with different VTLs are related by

$$\tilde{s}_n^A(i, \tau_i(p)) = \tilde{s}_n^B(i + \alpha_T, \tau_{i+\alpha_T}(p)), \tag{5.5}$$

where α_T is proportional to the ratio between the VTLs of A and B. Thus, a change of VTL leads to a shift of the formants along ridges that pass through the same cycles of the impulse responses of all subbands. In the works of Patterson et al. (2007) and Irino and Patterson (2002) the representation of the SAI in this scale-cycle space is called *size-shape image* (SSI). The SSI space is scale-shift covariant, which means that the effects due to different VTLs appear solely as translations along the subband axis.

Now, let $\boldsymbol{p} = (p_1, p_2, \ldots, p_M)$ contain cycle numbers. We define a monomial $\tilde{m}$ on base of the SAI space as

$$\tilde{m}(n; w, \boldsymbol{k}, \boldsymbol{l}, \boldsymbol{m}, \boldsymbol{p}) := \left[\prod_{i=1}^{M} \tilde{s}_{n+m_i}^{l_i}(k_i + w, \tau_{k_i+w}(p_i)) \right]^{1/\gamma(\boldsymbol{l})}. \tag{5.6}$$

With the definition from Equation (5.6), a feature component $A_{\tilde{m}}(n)$ is then computed as defined in Equation (5.2). The features based on the SAI will be referred to as AIM-IIFs in the following. We have used linear interpolation in this work to compute $\tilde{s}_n(i, \tau_i(p))$.

The idea of applying an invariant transform on the SAI was also part in the work of Irino and Patterson (2002), where an adapted form of the Mellin transform was used to compute a VTL-invariant representation of speech signals: The Mellin image (Irino and Patterson, 2002) is essentially the magnitude of the Fourier transform of the corresponding SSI vector and, thus, is also translation invariant. However, compared to the approach proposed in this work, the Mellin image has at least two disadvantages: Though invariant to translations, the magnitude of the Fourier transform is also invariant to additional transformations like mirroring. Also, the data rate of the Mellin image is as high as the one of the SAI and, thus, would need to be reduced prior to be fed into an ASR system. A benefit of the IIFs is that only selected segments of constant cycle numbers need to be considered, so that a transformation of the complete SAI into an SSI is not necessary. Furthermore, the extraction of AIM-IIFs also leads to a significant reduction of the data rate that is comparable to that of cepstral features.

5.3.3. Experimental Setup and Baselines

The following recognition experiments have been conducted on the Aurora-2 task. The standard training and test sets as they are published together with the corpus data (see also Appendix B.2) were taken here. These include utterances with *signal-to-noise ratios* (SNR) of 20, 15, 10, 5, 0, and -5 dB. Both clean speech and multi-condition training were considered. Average accuracies of all three test sets are shown in the following. Throughout the experiments, the same HTK-based back-end was used. Whole-word left-to-right models with 11 to 17 states depending on the average utterance lengths of the digits were used. Four Gaussians were used in the mixtures of the individual states, and the covariance matrices were constrained to be of diagonal form. All features were extracted with a frame rate of 100 Hz. First- and second-order time-derivative approximations were appended to all feature vectors. In case of integration features, a *linear discriminant analysis* (LDA) with a target dimensionality of 55 followed by a *maximum-likelihood linear transform* (MLLT) (see Section 2.3) was computed to reduce the feature vector dimensionality. The target dimensionality of 55 was empirically chosen in preliminary experiments.

Baseline accuracies were generated with MFCCs using the standard setup of HTK together with cepstral mean subtraction and also with *power-normalized cepstral*

Table 5.4.: Recognition accuracies [%] for IIFs, PN-IIFs, and AIM-IIFs for clean and multi-condition training on Aurora-2.

SNR	clean			multi-condition		
(dB)	IIF	IIF_{PN}	IIF_{AIM}	IIF	IIF_{PN}	IIF_{AIM}
∞	99.2	98.8	98.5	98.7	98.9	98.5
20	97.8	97.8	97.4	98.3	98.5	98.2
15	95.4	96.0	95.2	97.4	97.9	97.6
10	89.5	91.0	88.0	95.6	96.0	95.7
5	72.8	77.7	70.1	89.2	90.3	88.9
0	41.0	50.4	40.5	68.1	72.9	67.4
−5	14.0	22.1	17.3	31.0	39.1	29.6
Avg.	72.8	76.2	72.4	82.6	84.8	82.3

coefficients (PNCC) (see Section 2.2.6) as second feature type. PNCCs are cepstral coefficient-based features that can efficiently be computed and have recently shown a comparable noise-robustness in comparison to feature enhancement methods like Vector-Taylor series expansion or SPLICE. The results of these two feature types for both training modes were already shown in Table 5.1 on page 124.

The results clearly show the advantage of PNCCs compared to MFCCs under noisy conditions. While performing similar in case of clean speech, the PNCC-based ASR system achieves accuracies that are increasingly better in terms of accuracy the lower the SNR becomes. This holds for clean-speech training as well as for multi-condition training and supports the results from Kim and Stern (2010a). To allow for a comparison with integration-based features, accuracies of IIFs (Müller and Mertins, 2011a) and PN-IIFs (Müller and Mertins, 2011b) are shown in Table 5.4. PN-IIFs combine the methods for increasing the noise robustness of the PNCCs with invariant integration, which further increases the robustness to the effects of VTL differences. Table 5.4 also shows the results for the AIM-IIFs. For the feature selection, the same method as for the "standard" IIFs was used. We selected sets of 30 AIM-IIFs with the constraint of using only at most a cycle number of three. This constraint was introduced, because the glottal pulse rate imposes an upper limit for the time-interval before resonance and pulse information are superimposed in this space (Patterson et al., 2007; Irino and Patterson, 2002).

The IIF-based ASR system achieves accuracies that are higher than the MFCC-based system for all SNRs and for both training conditions. In comparison to PNCCs, however, they do not perform as well, which was also observed in the work of Müller and Mertins (2011b) and motivated PN-IIFs. PN-IIFs perform better than IIFs under all SNRs and better than PNCCs for SNRs down to 5 dB for both

Table 5.5.: Accuracies [%] for AIM-IIFs combined with originally proposed IIFs, PNCCs, and with PN-IIFs for clean and multi-condition training on Aurora-2.

	clean			multi-condition		
SNR (dB)	**IIF_{AIM} +IIF**	**+PNCC**	**+IIF_{PN}**	**IIF_{AIM} +IIF**	**+PNCC**	**+IIF_{PN}**
∞	99.2	99.0	99.0	98.8	98.7	98.9
20	97.9	97.8	98.2	98.2	98.2	98.5
15	96.0	96.5	96.7	97.4	97.6	98.0
10	90.9	92.3	92.6	95.5	96.2	96.4
5	77.3	81.1	81.3	89.8	91.3	91.6
0	48.4	54.7	57.0	71.9	75.5	76.4
−5	17.9	21.7	26.3	33.9	39.8	43.4
Avg.	75.4	77.6	**78.7**	83.7	85.3	**86.2**

training modes. A reason for the abrupt decrease of accuracy for lower SNRs in case of the PN-IIFs might be the fact that the reduction matrix, which is estimated with LDA on the training data, does not generalize well for noise scenarios with very low SNRs. Also, the parameters chosen for the power-normalization feature-enhancement stage might still not be optimal. A comparable performance of these feature types was also observed on artificially distorted TIMIT speech signals as presented in the work of Müller and Mertins (2011b).

5.3.4. Results for AIM-IIFs and Feature Combinations

The AIM-IIF-based ASR system shows a comparable performance to the IIF-based system for both training conditions in Table 5.1. Walters (2011) observed that SAI-based features yield a higher robustness under noisy conditions compared to MFCCs, while performing worse under clean conditions. This cannot be observed for the results in Table 5.4. However, due to the different way of processing the speech signal in comparison to the standard filter banks, we assume that SAI-based signal representations are prone to different kinds of errors. Thus, we investigated if a combination of AIM-IIF vectors together with IIF, PNCC, or PN-IIF feature vectors yields an enhancement in accuracy. In contrast to ROVER, this approach has the advantage that only a single ASR system is used. Again, we used LDA and MLLT to reduce the resulting feature dimension down to 55 and to decorrelate the features. The results of these experiments are shown in Table 5.5. It can be observed that the concatenation of all three feature types with AIM-IIFs generally increases the accuracy. Combining AIM-IIFs and PN-IIFs into one feature vector yields the highest accuracies within the experiments of this work and gives average accuracies (over all SNRs) of 78.7% and 86.2% for clean speech and multi-condition training, respectively.

6 Estimating the Spectral Effects due to Vocal-Tract Length Changes

For an effective feature-space normalization method that targets the vocal tract length as variability, a model is needed that approximates the effects due to different vocal tract lengths (VTL) sufficiently well. With respect to the time-frequency domain these effects are commonly described by one-parametric warping functions as described in more detail in Section 2.5. A piecewise-linear warping function that is applied on whole utterances is among the most commonly used functions for VTLN. Even though, methods were proposed that refine the standard VTLN procedure either by using more than one warping function for a single utterance, or by increasing the search space of the Viterbi decoder (see Section 2.5.1 on page 52f), the underlying principles of these methods still model the warping according to the assumptions of a lossless, uniform tube. In this chapter, two approaches are described that try to estimate nonparametric warping functions and that describe the spectral effects due to different VTLs without the model of a lossless, uniform tube. While the first approach is data-driven and makes use of a training set, the second approach is model-driven and relies on a vocal-tract model that is usually used for articulatory speech synthesis.

The first section of this chapter describes the data-driven approach and describes the used method for relating spectral profiles to each other in terms of displacement fields. The second section introduces a lossy wave-reflection model for the simulation of the vocal tract. This model is then used within the presented model-driven method for estimating the spectral effects due to different VTLs.

6.1. Data-driven Estimation of Spectral Effects due to Vocal-Tract Length Changes

Speaker-normalization and -adaptation methods are commonly used in speaker-independent automatic speech recognition (ASR) systems to handle inter-speaker variability. While "speaker-adaptation" usually refers to an adaptation of the acoustic model parameters with a maximum-likelihood linear regression (MLLR) approach (Gales, 1998), the term "speaker-normalization" is mostly used in the context of vocal tract length normalization (VTLN) methods (Lee and Rose, 1996), which try to compensate for the effects of different vocal tract lengths (VTL) on the feature extraction stage. As described in Section 2.5.1, this compensation is working on the whole utterance by either warping the frequency centers of the used filter bank or by warping the frequency axis of the output of the filter bank. Assuming a lossless, uniform tube model of length l, the resonance frequencies F_i occur at $F_i = c/(4l)$, $i = 1, 2, 3, \ldots$, where c is the speed of sound. This linear scaling of the resonances for different tube lengths is the basis for the often used piecewise-linear warping function as described, for example, by Hain et al. (1999). Different types of other warping functions were analyzed (Uebel and Woodland, 1999), but did not show any significant advances with respect to accuracy compared to piecewise-linear warping.

In the following, a method is proposed that accounts for two additional factors that are not or only roughly accounted for in the commonly used VTLN approach: Usually, the whole utterance of a single speaker is warped with only a single warping factor. While this approach mitigates the average effect of different VTLs on a per-speaker basis, it does not consider the fact that the VTL of a single speaker changes when producing phonemes where, for example, the lips are lengthened or the larynx is lowered (Mathur et al., 2006). There are works that follow the idea of using more than one warping parameter for normalizing the TF representation of an utterance of a single speaker: Maragakis and Potamianos (2008) proposed a region-based VTLN approach where a parameter for a piecewise-linear warping function is estimated for up to five phoneme groups during decoding. A method for a frame-wise warping parameter estimation was proposed by Miguel et al. (2005), where the Viterbi search space is augmented with a search for an optimal warping parameter. Both methods use a piecewise-linear warping function.

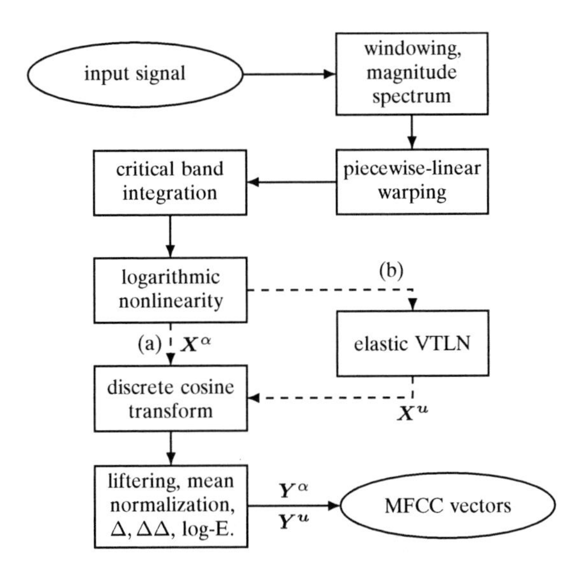

Figure 6.1.: *Computation of VTL normalized MFCC vectors (a) in its common form denoted as* $\boldsymbol{Y}^{\alpha}$*, (b) with the proposed enhancement denoted as* $\boldsymbol{Y}^{u}$.

Here, a data-driven method for refining the TF representation as output of the commonly used one-parameter VTLN approach is described. The proposed method makes use of elastic registration with specific constraints for the task of VTLN for ASR. The resulting warping functions are nonparametric and allow for a high degree of freedom.

The next section describes the idea of the proposed method and gives some details about its implementation. It was published by Müller and Mertins (2012a). Section 6.1.2 explains the experiments and analyzes the proposed method with respect to the resulting ASR performance. Conclusions are given in Section 6.1.4.

6.1.1. Vocal Tract Length Normalization and Elastic Registration

For the evaluation of the following method mel frequency cepstral coefficients (MFCC) are used. The procedure for the computation of MFCC vectors with an integrated (optional) warping of the frequency axis is illustrated in Figure 6.1. However, the proposed method can be used with any feature type with an intermediate spectral representation. An often used implementation of VTLN follows the procedures for speaker-adaptive training (SAT) and a two-pass decoding strategy as described by Welling et al. (2002). The method proposed in the following can be regarded as an additional feature enhancement step to standard VTLN.

Standard Vocal Tract Length Normalization

In this work, we consider a set of global warping factors $\alpha = \{-0.88, -0.9, \dots, 1.12\}$, where we refer to $\alpha_N = 1$ as the "neutral warping factor" in the following. The SAT procedure was already described in Section 2.5.1 and is repeated here for the introduction of certain mathematical notions. It can be formally summarized as follows: First, let $r = 1, \dots, R$ be utterance indices. Using the nonnormalized observations $\boldsymbol{Y}^{(r)}$ an acoustic model $\boldsymbol{\lambda}$ with single Gaussians per state is estimated,

$$\boldsymbol{\lambda} = \arg\max_{\widehat{\boldsymbol{\lambda}}} \prod_{r=1}^{R} p\left(\boldsymbol{Y}^{(r)} \mid \boldsymbol{W}^{(r)}; \widehat{\boldsymbol{\lambda}}\right). \tag{6.1}$$

Second, for each utterance the warping factor $\alpha^{(r)}$ is determined with the model $\boldsymbol{\lambda}$ and the ground-truth transcriptions $\boldsymbol{W}^{(r)}$ in a maximum likelihood sense,

$$\alpha^{(r)} = \arg\max_{\alpha} p(\boldsymbol{Y}^{(r),\alpha} \mid \boldsymbol{W}^{(r)}; \boldsymbol{\lambda}), \quad r = 1, \dots, R. \tag{6.2}$$

As third step, a VTL normalized acoustic model $\boldsymbol{\lambda}'$ is estimated using the normalized observations $\boldsymbol{Y}^{(r),\alpha^{(r)}}$ for each utterance r,

$$\boldsymbol{\lambda}' = \arg\max_{\widehat{\boldsymbol{\lambda}}} \prod_{r=1}^{R} p\left(\boldsymbol{Y}^{(r),\alpha^{(r)}} \mid \boldsymbol{W}^{(r)}; \widehat{\boldsymbol{\lambda}}\right). \tag{6.3}$$

For the recognition of a given observation sequence $\boldsymbol{Y}$ with the SAT acoustic model $\boldsymbol{\lambda}'$, a suboptimal two-pass strategy (Welling et al., 2002) can be applied as follows: A first decoding pass with nonnormalized observations $\boldsymbol{Y}$ and acoustic model $\boldsymbol{\lambda}$ yields a hypothesized transcription $\widetilde{\boldsymbol{W}}$,

$$\widetilde{\boldsymbol{W}} = \arg\max_{W} \{P(\boldsymbol{W}) \cdot p\,(\boldsymbol{Y} \mid \boldsymbol{W}; \boldsymbol{\lambda})\}. \tag{6.4}$$

Given the normalized model $\boldsymbol{\lambda}'$ and the hypothesis $\widetilde{\boldsymbol{W}}$, a warping factor $\widetilde{\alpha}$ is selected that yields the highest likelihood,

$$\widetilde{\alpha} = \arg\max_{\alpha} p\left(\boldsymbol{Y}^{\alpha} \mid \widetilde{\boldsymbol{W}}; \boldsymbol{\lambda}'\right). \tag{6.5}$$

A second decoding pass with normalized observations $\boldsymbol{Y}^{\widetilde{\alpha}}$ and normalized model $\boldsymbol{\lambda}'$ yields the final transcription,

$$\arg\max_{\boldsymbol{W}} \left\{P(\boldsymbol{W}) \cdot p\left(\boldsymbol{Y}^{\widetilde{\alpha}} \mid \boldsymbol{W}; \boldsymbol{\lambda}'\right)\right\}. \tag{6.6}$$

Elastic Vocal Tract Length Normalization The standard VTLN approach as described above can be seen as trying to deform the magnitude spectrum such that the deformed spectrum is more similar to a corresponding spectrum that would have been generated by a speaker associated with a neutral warping factor. Ideally, the deformation is context-dependent and has a high degree of freedom, which allows for the modeling of a wide range of spectral effects due to different VTLs.

Let us assume we have filter bank outputs $\boldsymbol{X}^{\alpha}$ that have been normalized with the VTLN approach as summarized in Section 6.1.1 and let $\boldsymbol{g} = (g_1, g_2, \ldots, g_G)$ denote the indices of utterances that are associated with the neutral warping parameter α_N. Furthermore, let $\boldsymbol{\Lambda}$ be a GMM based acoustic model whose parameters have been trained on the normalized outputs $\boldsymbol{X}^{\alpha}$ that are associated with the neutral warping parameter α_N,

$$\boldsymbol{\Lambda} = \arg\max_{\widehat{\boldsymbol{\Lambda}}} \prod_{k=1}^{G} p\left(\boldsymbol{X}^{(g_k),\alpha_N} \mid \boldsymbol{W}^{(g_k)}; \widehat{\boldsymbol{\Lambda}}\right). \tag{6.7}$$

Due to the GMM (here with M Gaussians) the probability density function (PDF) modeled by a single state j of an acoustic model is given as formulated in Equation (2.43) and is repeated here for convinience,

$$b_j(\boldsymbol{x}_t) = \sum_{m=1}^{M} c^{(jm)} \mathcal{N}\left(\boldsymbol{x}_t;\, \boldsymbol{\mu}^{(jm)},\, \boldsymbol{\Sigma}^{(jm)}\right). \tag{6.8}$$

In Equation (6.8) $\boldsymbol{x}_t$ is a single observation vector, $c^{(jm)}$ is a weighting coefficient, and $\mathcal{N}(\,\cdot\,;\, \boldsymbol{\mu},\, \boldsymbol{\Sigma})$ is a multivariate Gaussian PDF with mean $\boldsymbol{\mu}$ and covariance $\boldsymbol{\Sigma}$,

$$\mathcal{N}(\boldsymbol{x};\, \boldsymbol{\mu},\, \boldsymbol{\Sigma}) = \frac{1}{\sqrt{(2\pi)^n \left|\boldsymbol{\Sigma}\right|}} e^{-\frac{1}{2}(\boldsymbol{x}-\boldsymbol{\mu})^T \boldsymbol{\Sigma}^{-1} (\boldsymbol{x}-\boldsymbol{\mu})}. \tag{6.9}$$

Obviously, the likelihood $b_j(\boldsymbol{x}_t)$ in Equation (6.8) can be maximized with

$$\boldsymbol{x}^{(j)} = \arg\max_{\widehat{\boldsymbol{x}}_t} b_j(\widehat{\boldsymbol{x}}_t) = \sum_{m=1}^{M} c^{(jm)} \boldsymbol{\mu}^{(jm)}, \tag{6.10}$$

and it can be seen that the maximum can be determined if the state j is known. Now, let $S_r(\boldsymbol{X}, \boldsymbol{\lambda}', \boldsymbol{W}) = (S_1, S_2, \ldots, S_T)$ denote the state sequence of utterance r that is estimated with forced-alignment based on a feature vector sequence $\boldsymbol{X}$,

$$\boldsymbol{X} = \begin{bmatrix} \boldsymbol{x}_1 & \boldsymbol{x}_2 & \ldots & \boldsymbol{x}_T \end{bmatrix}, \tag{6.11}$$

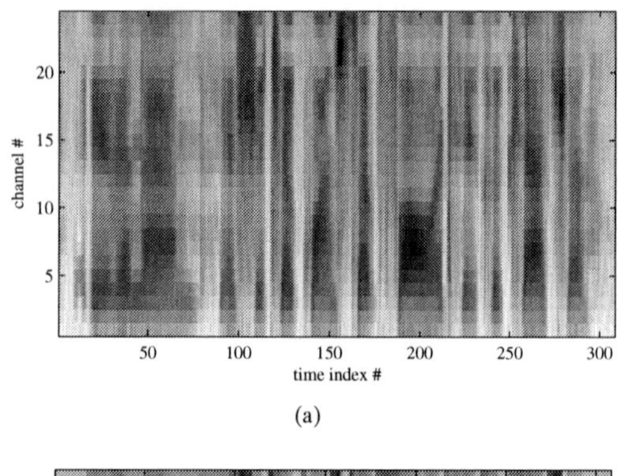

(a)

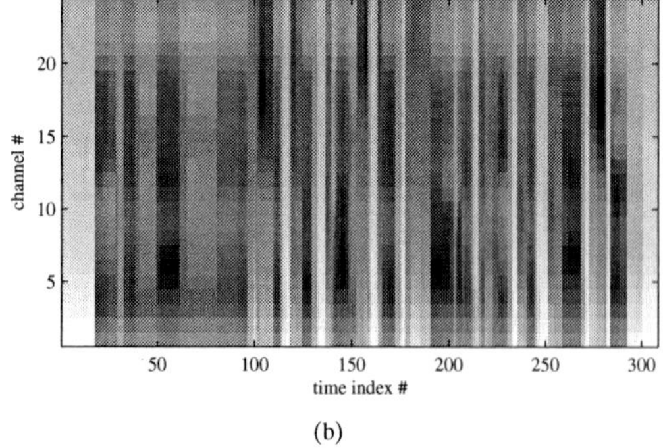

(b)

Figure 6.2.: (a) Original observation sequence $\boldsymbol{X}$, (b) optimal observation sequence $\boldsymbol{X}^*$.

an acoustic model $\boldsymbol{\lambda}'$, and a given transcription $\boldsymbol{W}$. The acoustic likelihood for $\boldsymbol{X}$ and S_r given $\boldsymbol{\Lambda}$ is

$$p(\boldsymbol{X}, S_r \mid \boldsymbol{\Lambda}) = \prod_{t=1}^{T} b_{S_t}(\boldsymbol{x}_t). \tag{6.12}$$

Equation (6.12) would be maximized with

$$\boldsymbol{X}^* = \begin{bmatrix} \boldsymbol{x}_1^* & \boldsymbol{x}_2^* & \dots & \boldsymbol{x}_T^* \end{bmatrix} \quad \text{where} \quad \boldsymbol{x}_t^* = \boldsymbol{x}^{(S_t)}. \tag{6.13}$$

Figure 6.2 (a) shows an exemplary filter bank output $\boldsymbol{X}$ of a single utterance. Using a three-state left-to-right monophone model $\boldsymbol{\lambda}'$, a forced-alignment $\boldsymbol{W}$ was estimated, which yields a state-sequence $S(\boldsymbol{X}, \boldsymbol{\lambda}', \boldsymbol{W})$. The optimal feature vector sequence $\boldsymbol{X}^*$ according to Equation (6.13) is shown in Figure 6.2 (b).

We want to describe the spectral effects due to VTL changes for each frame of a whole utterance. The key idea of the following method is to find a transformation such that a transformed observation sequence is similar to its optimal feature vector sequence. This procedure is called "registration" and is actively researched within the field of image processing. As is described in more detail in the following, the objective function to be optimized contains a term that is based on the linearized elastic potential. Therefore, the proposed method is referred to as "elastic VTLN".

Elastic Registration Details to the following introduction about the applied registration approach can be found in the work of Modersitzki (2004). In general, the goal of registration can be stated as follows: Given a reference $\boldsymbol{R}$ and template $\boldsymbol{T}$ and a mapping $\boldsymbol{R}, \boldsymbol{T} : \mathbb{R}^2 \rightarrow \mathbb{R}$, we want to find a displacement $\boldsymbol{u} : \mathbb{R}^2 \rightarrow \mathbb{R}^2$, such that the transformed template $\boldsymbol{T}^{\boldsymbol{u}}$,

$$\boldsymbol{T}^{\boldsymbol{u}} := \boldsymbol{T}(x - \boldsymbol{u}(x)), \tag{6.14}$$

is similar to $\boldsymbol{R}$. For the computation of $\boldsymbol{T}^{\boldsymbol{u}}$ a linear interpolation scheme is used here and the boundaries of $\boldsymbol{T}$ were extended with linear regression. The similarity is quantified with a distance measure $\mathcal{D}[\boldsymbol{R}, \boldsymbol{T}^{\boldsymbol{u}}] : \mathbb{R}^2 \rightarrow \mathbb{R}$. By introducing a regularization term $\mathcal{S}[\boldsymbol{u}] : \mathbb{R}^2 \rightarrow \mathbb{R}$ prior knowledge can be introduced and the numerical solution becomes more stable. The constrained optimization problem then reads

$$\min_{\boldsymbol{u}} \mathcal{D}[\boldsymbol{R}, \boldsymbol{T}^{\boldsymbol{u}}] + \nu \mathcal{S}[\boldsymbol{u}] \quad \text{subject to} \quad \boldsymbol{u} \in \mathcal{M}, \tag{6.15}$$

where $\nu \in \mathbb{R}^+$ is a regularization parameter, and $\mathcal{M}$ is a set of admissible transformations.

As distance measure $\mathcal{D}$ the correlation-based distance measure (Modersitzki, 2004) is used,

$$\mathcal{D}^{\text{corr}}[\boldsymbol{R}, \boldsymbol{T}^{\boldsymbol{u}}] = \left\langle \frac{\boldsymbol{R} - \mu(\boldsymbol{R})}{\sigma(\boldsymbol{R})}, \frac{\boldsymbol{T}^{\boldsymbol{u}} - \mu(\boldsymbol{T}^{\boldsymbol{u}})}{\sigma(\boldsymbol{T}^{\boldsymbol{u}})} \right\rangle_{L_2}, \tag{6.16}$$

where $\mu(\cdot)$ and $\sigma(\cdot)$ denote the mean and standard deviation, respectively. The choice for the regularizer in this work can be motivated, for example, by considering the spectral effects of spatially restricted VTL changes. By means of an articulatory speech synthesis model it is shown by Mathur et al. (2006) that an elongation at the lips, the larynx, or a mid segment yield a warping of resonance frequencies that is not linear with frequency. In the two-dimensional case the elastic regularizer $\mathcal{S}^{\text{elast}}$ (Modersitzki, 2004) can be seen as a rubber foil that induces tension if

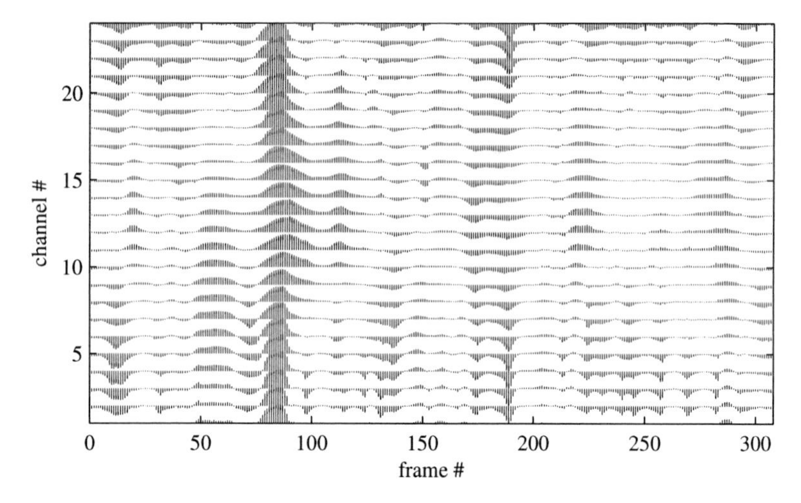

Figure 6.3.: Exemplary displacement field $\boldsymbol{u}$.

deformed. For two dimensions it is defined as

$$\mathcal{S}^{\mathrm{elast}}[\boldsymbol{u}] = \frac{1}{2} \int_{\Omega} \sum_{d=1}^{2} \rho \|\nabla \boldsymbol{u}_d\|^2 + (\rho + \kappa)(\operatorname{div} \boldsymbol{u})^2 \, \mathrm{d}x, \tag{6.17}$$

where κ, $\rho \in \mathbb{R}^+$ are the so-called Navier-Lamé constants, which control the elastic behavior of the deformation, ∇ denotes a gradient, and div is the divergence operator.

For the optimization of Equation (6.15) we use the first-optimize-then-discretize approach. That means, a minimizer of the objective function is determined first, which leads to a nonlinear system of partial differential equations (PDE). Then, the PDE is discretized and solved with a fixed-point iteration scheme here. To constrain the possible solutions with displacements along the subband axis, the displacements that occur along the time axis are set to zero in each iteration of the numerical solution while keeping the displacements along the subband axis. The choice of the Navier-Lamé constants is highly task dependent. In this work they were set to $\kappa = 0$ and $\rho = 1$, which is a common choice (Modersitzki, 2004).

As an example, a displacement field for the reference and template signals shown in Figure 6.2 (b) and (a), respectively, can be seen in Figure 6.3. The displacements along the subband axis for each component are clearly visible. The chosen regularization parameter yields spatially restricted and smooth displacements.

Using Elastic Registration for VTLN: Elastic VTLN The standard VTLN approach can be used for SAT, as well as for VTL normalization during recognition. By making use of elastic VTLN, we propose procedures for both cases to enhance the overall performance of the ASR system in the following.

Starting with a SAT acoustic model $\boldsymbol{\lambda}'$, the following method aims to further decrease the effects of inter-speaker-variabilities that result in translations along the subband axis. In a first step, an acoustic model $\boldsymbol{\Lambda}$ is trained only on utterances that are associated with the neutral warping parameter (see Section 6.1.1). With the ground-truth labels of the training data, a maximum-likelihood (ML) state alignment is computed. For each training observation sequence $\boldsymbol{X}^{(r)}$, an optimal observation sequence $\boldsymbol{X}^{(r),*}$ is generated and a displacement field $\boldsymbol{u}^{(r)}$ is estimated with $\boldsymbol{X}^{(r),*}$ being the reference and $\boldsymbol{X}^{(r)}$ being the template,

$$\boldsymbol{u}^{(r)} = \arg\min_{\widehat{\boldsymbol{u}}} \ \mathcal{D}^{\text{corr}}\left[\boldsymbol{X}^{(r),*}, \boldsymbol{X}^{(r),\widehat{\boldsymbol{u}}}\right] + \nu\mathcal{S}^{\text{elast}}\left[\widehat{\boldsymbol{u}}\right]. \tag{6.18}$$

The application of the displacements for each utterance yields a warped spectral representation $\boldsymbol{X}^{(r),\boldsymbol{u}}$. A subsequent computation of cepstral-coefficient based features on the basis of the warped representations (see also Figure 6.1) yields the final observations $\boldsymbol{Y}^{(r),\boldsymbol{u}}$. These are used for a re-estimation of the acoustic model parameters, which leads to the final acoustic model $\boldsymbol{\lambda}''$,

$$\boldsymbol{\lambda}'' = \arg\max_{\widehat{\boldsymbol{\lambda}}} \prod_{r=1}^{R} p\left(\boldsymbol{Y}^{(r),\boldsymbol{u}} \mid \boldsymbol{W}^{(r)}, \widehat{\boldsymbol{\lambda}}\right). \tag{6.19}$$

Similar to the standard VTLN approach, the decoding of features with elastic VTLN uses the hypothesis $\widetilde{\boldsymbol{W}}$ from a first decoding pass for an ML state-alignment. The output of the state-alignment is used to generate a hypothetically optimal observation sequence $\widetilde{\boldsymbol{X}}^{*}$, which, in turn, is used as reference for a subsequent elastic registration. The resulting displacement $\boldsymbol{u}$ is then used to compute a deformed spectral representation $\boldsymbol{X}^{\boldsymbol{u}}$. The deformed spectral values are used for the extraction of cepstral-coefficient based features $\boldsymbol{Y}^{\boldsymbol{u}}$. A second decoding pass yields the final transcription.

6.1.2. Experiments

The TIMIT corpus with its standard training and test sets (without SA sentences) was used throughout the experiments. The training set consists of 3696 utterances from 462 different speakers. The test set consists of 1344 utterances from another 168 different speakers. Following the standard procedure for TIMIT, the

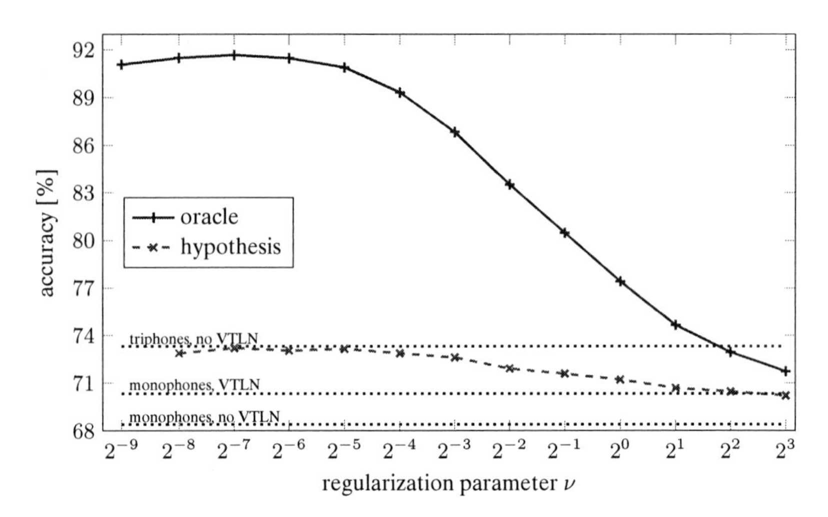

Figure 6.4.: Resulting recognition accuracies using elastic VTLN for oracle (solid) and hypothetical (dashed) transcriptions. The baseline accuracies obtained with the monophone and triphone systems (with and without standard VTLN) are shown as dotted lines.

initial phoneme set was folded to 48 phonemes. Three-state left-to-right monophone models with up to 16 Gaussians and diagonal covariance matrices together with bigram statistics were used. For the computation of the recognition accuracy, the transcriptions were further folded to 39 phonemes. Monophone models were used here to decrease the computational load, thus, making the analysis of the proposed elastic VTLN method more feasible. The feature extraction follows the procedure as depicted in Figure 6.1 and yields 39 dimensional MFCC vectors. The baseline recognition accuracy of the system without VTLN is 68.4%, and 70.3% with standard VTLN. As additional baseline, triphone modeling with rule-based state-clustering yields an accuracy of 73.3% without VTLN, and 74.5% with standard VTLN.

In a first step, an upper bound for the accuracy obtained with elastic VTLN was determined. This was done by estimating state alignments based on oracle transcriptions for both the training as well as for the test utterances. Features were computed with the resulting deformations as described in Section 6.1.1 and recognitions experiments were conducted with the monophone system. The accuracies for different choices of the regularization weights ν are shown in Figure 6.4 as solid line. The impact of a large regularization coefficient is clearly visible: The larger the weight, the smaller are the resulting displacements towards optimal spectral representations. An optimal choice for ν with respect to the accuracy is given by $\nu = 0.008$ in these experiments. As is described next, this holds for the use of both the oracle as well as the hypothesized transcription.

The potential of elastic VTLN is clearly shown with the accuracy reaching 91.7% with the oracle-transcription based monophone system. However, in practice, a hypothesized transcription from the first decoding pass has to be used for the normalization. To see how elastic VTLN performs under practical conditions, hypothesized transcriptions as output of the standard VTLN approach were used for the computation of the displacement fields in a second experiment. The results are shown in Figure 6.4 as dashed line with "x"-markers. It can be seen that a large regularization weight yields no performance improvements in comparison to standard VTLN. However, when choosing $\nu = 0.008$, the accuracy of the monophone system can be increased by more than four percentage points, reaching the accuracy of the triphone system.

6.1.3. Estimation of Global Warping Functions

The resulting displacements from the registration process as described in Section 6.1.1 can be used to compute transforms in a maximum a-posteriori sense that can be applied on whole utterances as it is done with the standard VTLN approach (see Section 2.5.1). While the use of more than a single warping function for a single utterance has proven to be able to improve the performance of ASR systems (as discussed in Section 6.1 and showed by the experiments above), a comparison between the accuracies based on the model-driven piecewise-linear warping function and a data-driven nonparametric warping function is interesting to see. In the following the computation of the global warping functions based on the estimation displacements as well as the corresponding phoneme recognition experiments are described.

In a first step the speakers of a training set have to be clustered according to their VTL. This could be done, for example, with the standard VTLN method that commonly considers between 13 and 19 warping parameters. In the following let the warping parameter $\alpha^{(r)}$ for each utterance r of a training set and the a-priories $P(S)$ for each phoneme state S be given. Furthermore, $\boldsymbol{u}_S^{\alpha_i}$ denotes the set of deformations that are assigned to warping factor α_i and to state S (obtained by forced-alignment). Then, the MAP deformations $\boldsymbol{u}^{\alpha_i}$ for each warping class α_i can

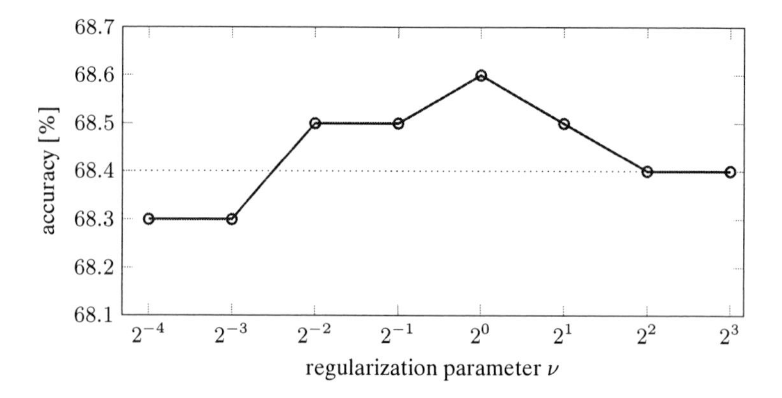

Figure 6.5.: Resulting recognition accuracies using MAP estimated global transforms for VTLN. The accuracy obtained with standard VTLN is indicated as dashed line.

be computed as weighted averages of the obtained deformation fields,

$$\boldsymbol{u}^{\alpha_i} = \sum_S P(S) \cdot E\{\boldsymbol{u}_S^{\alpha_i}\}. \tag{6.20}$$

This procedure yields a set of warping functions that can be used as replacement for the piecewise-linear warping function from the standard VTLN approach. The recognition accuracies obtained with this approach are illustrated in Figure 6.5. It can be seen that the obtained accuracy is only slightly higher than that obtained by standard VTLN. This indicates the importance of a context-dependent application of VTLN.

6.1.4. Discussion and Conclusions

We presented a method termed "elastic VTLN" for enhancing the VTLN approach. The method is data-driven and makes use of elastic registration with nonparametric deformations as output. Using elastic VTLN, the results show that it is possible to enhance the performance of a monophone system such that it reaches that of a triphone system.

6.2. Model-driven Estimation of Spectral Effects due to Vocal-Tract Length Changes

While the estimation of the spectral effects due to VTL changes in Section 6.1 was data-driven, a model-based approach is investigated in this section. Therefore, a so-called lossy wave-reflection analog model that was originally developed for articulatory speech synthesis is used for simulating the vocal tract. The implemented model is based on the works by Story (1995, 2005); Titze (2006), which, in turn, make use of the Kelly-Lochbaum model (K-L model, Kelly and Lochbaum, 1963). In the following a brief overview of the principles of the implemented model is given. For details the reader is conferred to the works of Liljencrants (1985); Story (1995, 2005); Titze (2006).

6.2.1. A Lossy Wave-Reflection Model for Simulating the Vocal Tract

Basic Scattering Equations The K-L model simulates the wave propagation within the vocal tract by representing the tract as a sequence of K uniform tubes of equal length Δx,

$$\Delta x = c/(2F_s), \tag{6.21}$$

where c is the speed of sound, F_s is the sampling rate, and different areas A_k, $1 \leq k \leq K$. The simulation begins with the lossless one-dimensional *wave equation*,

$$\frac{1}{c^2}\frac{\partial^2 P_k}{\partial t^2} = \frac{\partial^2 P_k}{\partial x^2}, \tag{6.22}$$

where P_k is the sound pressure in tube k, the displacement along the vocal tract is x, and t is the continuous time variable. A solution to Equation (6.22) is of the d'Alembert type (Titze, 2006),

$$P_k(x,t) = P_k^+(x-ct) + P_k^-(x+ct), \tag{6.23}$$

where the + and - superscripts denote traveling waves in forward and backward direction. Here, "forward" describes the direction of raising indices k. Following the simplified notation from Story (1995), forward and backward traveling waves of tube k are denoted in the following as

$$F_k = P_k^+(x-ct), \tag{6.24}$$

$$B_k = P_k^-(x+ct), \tag{6.25}$$

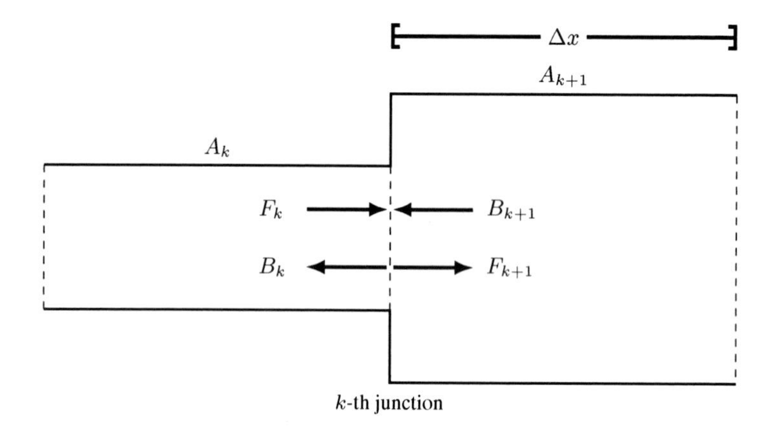

Figure 6.6.: Tube junction of one dimensional wave-reflection analog with wave scattering. The length of a single tubelet is indicated as Δx. When the forward traveling wave F_k and the backward traveling wave B_{k+1} meet at the k-th junction, they interact and result in the partial waves B_k and F_{k+1}.

so that the total pressure P_k in tube section k is

$$P_k = F_k + B_k \,. \tag{6.26}$$

Basically, Equation (6.23) states that the pressure in each tube at time instance t can be seen as one partial wave that travels forward to the next tube and one partial wave that travels backward to the previous tube. Figure 6.6 illustrates the following formulations. Before the partial waves travel across a joint they have the pressures F_k and B_{k+1}. When these waves meet at the junction they interact and result in the partial waves F_{k+1} and B_k, which leave the junction.

The flow U_k through a section k is the difference of the partial pressures F_k and B_k in relation to the *acoustic impedance* Z_k of that section,

$$U_k = \frac{1}{Z_k}(F_k - B_k)\,, \tag{6.27}$$

$$Z_k = \frac{\rho c}{A_k}\,, \tag{6.28}$$

where ρ is the density of air, and A_k is the area of section k. With the assumption that the pressure and the flow are continuous across section junctions, the partial pressures in the tubes k and $k+1$ are related by

$$F_k + B_k = F_{k+1} + B_{k+1}\,, \tag{6.29}$$

$$\frac{1}{Z_k}(F_k - B_k) = \frac{1}{Z_{k+1}}(F_{k+1} - B_{k+1})\,. \tag{6.30}$$

With respect to the k-th junction (between section k and $k+1$), the incoming pressures are F_k and B_{k+1}. The outgoing pressures of junction k are F_{k+1} and B_k. For the intended simulation of the wave propagation we are interested in the computation of the outgoing pressures given the incoming pressures. By using Equations (6.28), (6.29), and (6.30), and solving for F_{k+1} and B_k yields

$$F_{k+1} = F_k\left(\frac{2A_k}{A_k + A_{k+1}}\right) + B_{k+1}\left(\frac{A_{k+1} - A_k}{A_k + A_{k+1}}\right), \tag{6.31}$$

$$B_k = F_k\left(\frac{A_k - A_{k+1}}{A_k + A_{k+1}}\right) + B_{k+1}\left(\frac{2A_{k+1}}{A_k + A_{k+1}}\right). \tag{6.32}$$

By defining a *reflection coefficient* r_k that relates the areas of adjacent tubes to each other,

$$r_k = \frac{A_k - A_{k+1}}{A_k + A_{k+1}}\,, \tag{6.33}$$

the expressions for F_{k+1} and B_k can be simplified to

$$\begin{aligned} F_{k+1} &= (1 + r_k)F_k - \qquad r_k\,B_{k+1}\,, \\ B_k &= \qquad r_k\,F_k + (1 - r_k)B_{k+1}\,. \end{aligned} \tag{6.34}$$

A "one multiplier" lattice can be obtained with some rearrangement,

$$\begin{aligned} F_{k+1} &= s_k + F_k\,, \\ B_k &= s_k + B_{k+1}\,, \end{aligned} \tag{6.35}$$

where $s_k = r_k(F_k - B_{k+1})$. Equations (6.34) and (6.35), respectively, represent the *scattering equations* for simulating an ideal lossless, hard-walled vocal tract with uniform tubes.

It follows from Equation (6.21) that each segment length corresponds to the delay of a half-time sample. Because of this, the computations for one sample are arranged into two passes (Liljencrants, 1985): The first pass handles the even numbered joints and computes the odd numbered partial waves with forward direction and the even numbered backward waves. The second pass iterates over the odd joints

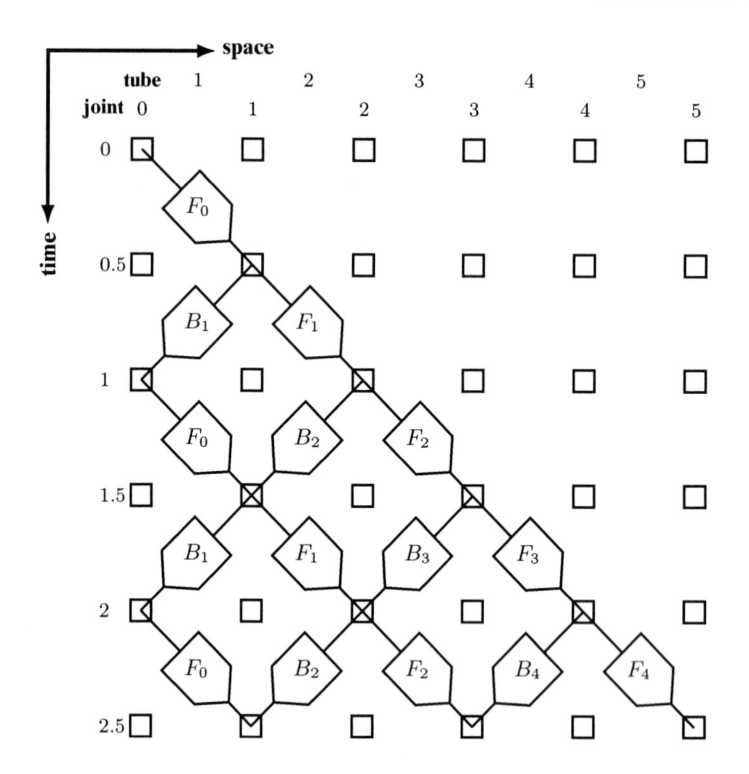

Figure 6.7.: Space time diagram of the computation sequence in the wave-reflection analog. Adapted from Liljencrants (1985).

and computes the even forward and the odd backward waves. The second step can be considered to be "half a sample later" and produces the partial waves incident on the even junctions for the next true time sample. Figure 6.7 shows all possible propagations of the partial wave components F_k and B_k as result of an impulse excitation as an adaptation from the corresponding illustration in the work of Liljencrants (1985).

The length of each tube corresponds to a travel time being only half of the sample interval. With a sampling rate $F_s = 44100$ Hz and assuming for the speed of sound $c = 35000$ cm/s, the length Δx of a single tube is $\Delta x = 0.3968$ cm. Thus, a vocal tract length of about 17.5 cm is simulated with a model made up of 44 tubes.

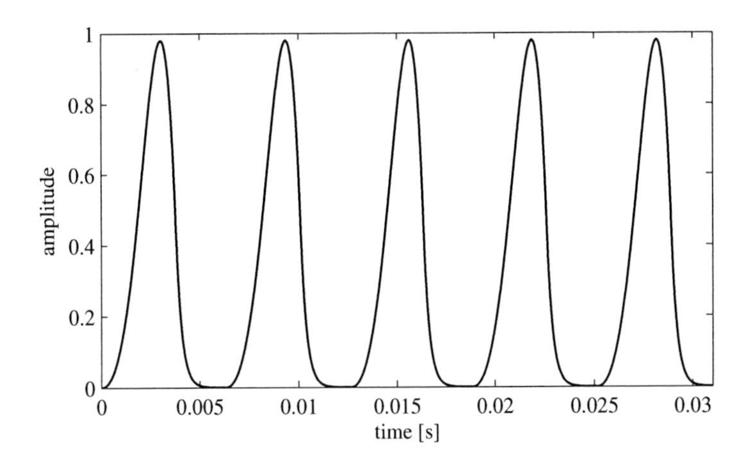

Figure 6.8.: Exemplary glottal waveform with a pitch of 160 Hz generated using the Liljencrants-Fant model (Fant et al., 1985).

Handling Glottal and Lip Boundaries Different types of voice sources have been proposed for this type of wave-reflection model (see, for example, Story, 1995, p. 80). They can be grouped into interactive and noninteractive sources. The former provides interaction between the tract and the glottal model, which, in turn, may even be connected to a subglottal model. The noninteractive sources inject a glottal waveform into the vocal tract model without any interaction between the tract and the voice source. For the synthesis of speech signals the glottal pulse model from Fant et al. (1985) as implemented by Brookes (2011) is used in this work. An exemplary glottal waveform with a pitch of 160 Hz is shown in Figure 6.8. The use of a pulse model allows a precise control of the pitch at the cost of a less natural sounding synthesis result. However, for most cases within this work only the impulse response of the vocal tract model is determined while the full synthesis of a speech signal was mainly used for a qualitative evaluation of the synthesis result. According to Titze (2006) with a given glottal flow U, the forward wave F_1 at the input of the vocal tract can be modeled with

$$F_1 = \frac{\rho c}{A_1} U + r_g B_1, \tag{6.36}$$

where r_g is the glottal reflection coefficient, commonly chosen as $r_g = 0.99$ in this noninteractive glottis-tract coupling. The transition from the last tube (representing the lips) to the open space leads to a change in acoustic impedance. The radiation impedance is approximated by a parallel resistive and inertive element (Titze, 2006,

p. 324) and yields linear filters for the partial waves in the last tube. The filters characteristics depend on the opening area of the last tube.

Simulation of Losses So far, the described model imitates a concatenation of ideal, uniform tubes without any losses. The simulation of different kinds of naturally occurring losses can be added to the model in a modular way and is described in the work of Story (1995). The kinds of losses described in that work are lumped element circuit approximations and involve

- yielding wall vibrations,
- viscous fluid losses,
- heat conduction losses,
- kinetic pressure drops, and
- sound radiation from skin surfaces.

Here, the implementation of the losses for this work are based on the descriptions given in the work of Story (1995). To verify the implementation comparable experiments with idealized vocal tract configurations as described by Story (1995) were conducted. For the verification of, for example, the yielding wall losses a uniform tube (consisting of 44 sections) with cross-sectional area equal to 4 cm^2 was used. With enabled radiation impedance at the lips, the frequency response for the hard walled tube is shown in Figure 6.9 (a) as solid line. Yielding walls should shift the first formant upward in frequency by about 30 to 40 Hz, while letting the higher formants nearly unaltered. The resulting frequency response with enabled yielding walls in the model is shown in Figure 6.9 (a) as dashed line. The expected frequency shifts are visible and reproduce the results described by Story (1995, p. 95). For the verification of the implementation of the viscous losses the cross-section area of the 44-segment tube was set to 0.1 cm^2 to increase the visibility of the loss effects. The frequency response for the tube with only the radiation impedance enabled is shown in Figure 6.9 (b) as solid line. The dashed line of Figure 6.9 (b) shows the frequency response with additional viscous losses enabled. As described by Story (1995, p. 111) it is expected that the effects of the viscous losses are greatest for frequencies of about 2 kHz and above and decrease the resonance frequencies with increasing frequency. These effects can be observed in Figure 6.9 (b).

Determining the Area Functions In order to produce different kinds of phones the areas of the individual tubes have to be set appropriately. A function that sets the individual tube areas with respect to the desired sound that the model

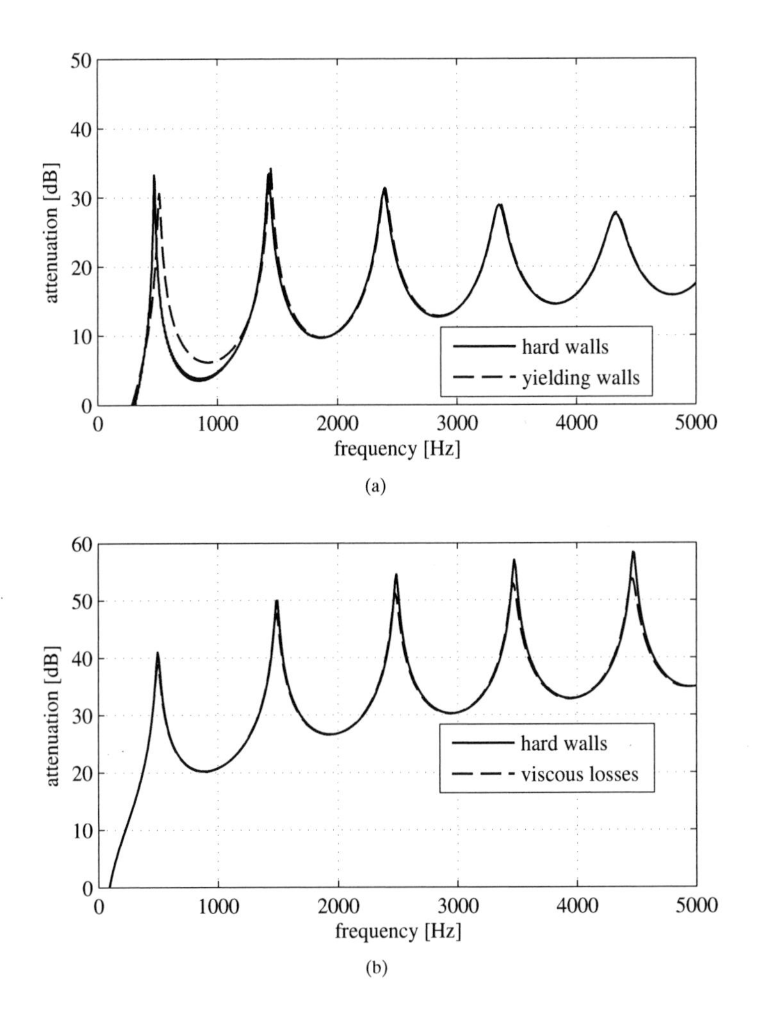

Figure 6.9.: Frequency responses of uniform tubes with cross-sectional area of (a) 4 cm^2, (b) 0.1 cm^2 without and with simulation of yielding walls and viscous losses, respectively.

should produce is called *area function* in the following. Because of the physiological correlation between the tube areas and the human vocal tract structure one way of determining the area functions is by means of *magnetic resonance imaging* (MRI) studies (for example, Story, 1995; Story et al., 1996; Kröger et al., 2000). Basically, the airway of the vocal tract is segmented within the three-dimensional data of the MRI and then the three-dimensional shape of the vocal tract is reconstructed. In a last step, the area function is determined by computing a three-dimensional centerline through the airway and determining the areas of the planes perpendicular to and located equally spaced on the centerline.

The area functions used here are based on the work described by Story (2005, and references therein). While it is possible to directly specify the individual areas with the results of the measured areas as described above, Story and Titze (1998) formulate the area functions in terms of a "neutral" diameter function $\Omega(k)$, $k = 1, \ldots, 44$, and proportional amounts of orthogonal basis functions Φ_1, Φ_2. In terms of the mean area function and the basis functions, Story (2005) defines the general area function $V(k,t)$ as

$$V(k,t) = \frac{\pi}{4} \left[\Omega(k) + q_1(t)\Phi_1(k) + q_2(t)\Phi_2(k)\right]^2 , \tag{6.37}$$

where the squaring and scaling operations convert the diameters to areas. The neutral diameter function $\Omega(k)$ and the basis functions $\Phi_1(k)$, $\Phi_2(k)$ are shown in Figure 6.10. Story (2005) provides mode coefficients for various vowels. However, the initial validation of the implementation used in this work did not reproduce the results published by Story (2005). In a personal correspondence with the author an error within the manuscript from Story (2005) was identified and the corrected mode coefficients were provided. These coefficients are listed in Table 6.1 and were used for the experiments in this work.

Possible Future Refinements for the Implementation The described model of the vocal tract so far allows for the simulation of different vocal tract configurations that capture the characteristics of various vowels. As it is described in the next section, this stage of modeling is sufficient for a model-driven estimation of spectral effects due to different vocal tract lengths. For an articulatory speech synthesis one would have to enhance the described model in various points in order to produce nonvowel sounds. For example, the nasal tract and its coupling with the vocal tract have been omitted so far. Approaches for the modeling of a nasal tract and its coupling to the vocal tract based on the K-L model are described by, for example, by Story (1995) and Titze (2006). A realistic glottal source for the production of voiced sounds would be another improvement to the model. Such a source model could be characterized by parameters with physiological analogies and an overview

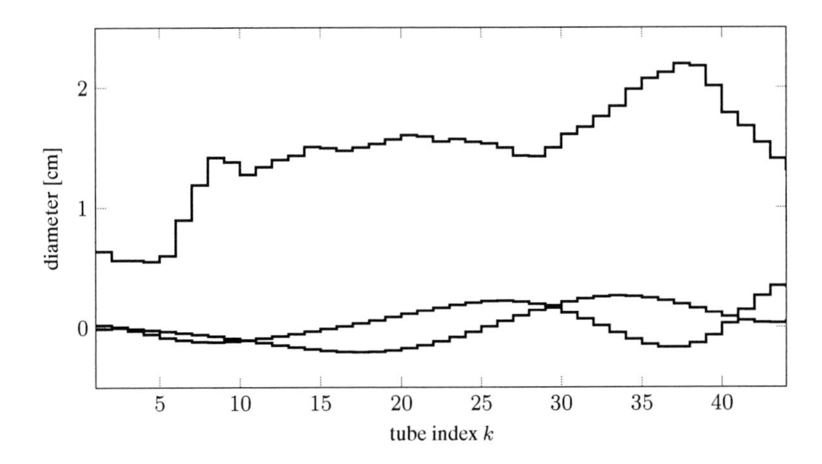

Figure 6.10.: Neutral diameter function $\Omega(k)$ and basis functions $\Phi_1(k)$, $\Phi_2(k)$ (Story, 2005), where the lowest tube index k refers to the boundary to the glottis and the highest index refers to the boundary to the open space.

of different approaches can be found in the book of Titze (2006). Another basic component for the synthesis of nonvowel sounds is the simulation of noise sound as they occur, for example, with fricative or stop sounds (Birkholz and Jackèl, 2006; Birkholz et al., 2007).

6.2.2. Estimating the Spectral Effects and Recognition Experiments

While the approach for estimating the relating transforms between the spectral envelopes of the same sounds originated from VTLs with different lengths in Section 6.1 is data-driven, this section describes a model-driven approach for such an estimation. Therefore, the vocal tract model, which was described in the previous section, is used to simulate different VTLs and the synthesized utterances are then passed to a front-end that extracts MFCC features. Similar to the data-driven approach from Section 6.1.1, the spectral profiles of the same phonème originated from vocal tract models with different lengths are then related with each other in terms of a displacement field with elastic registration.

The estimation of the warping functions is split into three stages: First, vowel sounds

Table 6.1.: Corrected mode coefficients q_1 and q_2 for tube model parametrization.

Vowel	q_1	q_2
i	-5.1758	0.9807
I	-2.5555	1.0428
eh	-1.0916	1.3310
ae	0.6768	2.3147
uh	2.5606	0.1282
a	3.8495	1.3562
aw	3.4713	-0.4930
U	1.7286	-1.9103
o	0.1016	-2.6849
u	-3.5654	-2.0655

are synthesized with the tract model. Nine VTLs linearly distributed within the range from 14.3 cm to 20.6 cm are simulated. Table 6.1 lists the parameters for ten vowel configurations. For each of these configurations, 121 variations were generated with additive differences Δ_p in the corresponding weighting coefficients for Φ_1, Φ_2, where

$$\Delta_p \in \{\pm 0.2, \pm 0.1, \pm 0.05, \pm 0.025, \pm 0.0125, 0\}. \tag{6.38}$$

This range of weighting deltas was determined empirically in preliminary experiments. For each tract configuration a 500 ms speech signal with a sampling rate of 16 kHz is generated and passed to a gammatone filter bank with 110 channels and with filter center frequencies ranging from 40 Hz to 8 kHz. Overall, 10890 utterances were synthesized. The spectral values from the filter bank output are averaged along the temporal axis and passed to a power-function nonlinearity with an exponent of 0.1 afterwards. In the second step displacements $\boldsymbol{u}$ for each utterance were computed with the elastic registration approach as described above. The compressed average spectral profiles of each pair of tract length and vowel type were concatenated and used as template $\boldsymbol{T}$. The corresponding spectral profiles of the model with "neutral" VTL of 17.5 cm are used as reference $\boldsymbol{R}$. Generally, the choice of the regularization parameter ν is task dependent and, thus, in the experiments ν was chosen from the range of $[2^{-5}, 2^{7}]$ for determining an optimal choice. The results of the second step are displacements for each considered tract configuration. Figure 6.11 shows the averaged final distance measure values for the different choices of ν relative to the distance measure based on the unwarped utterances. Overall, it can be seen that the choice of ν is proportional to the resulting distance measure value. For comparison with the piecewise-linear warping

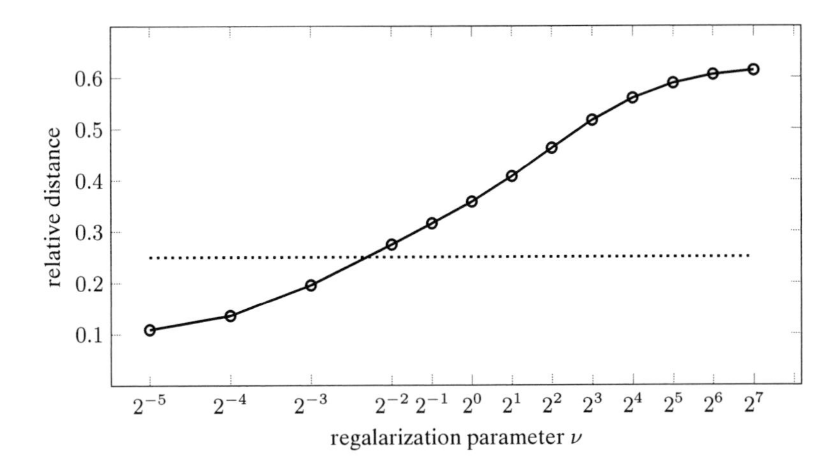

Figure 6.11.: Averaged final distance measure values for the different choices of ν relative to the distance measure based on the unwarped utterances (solid). Relative distance for standard VTLN with piecewise-linear warping function (dotted).

function that is used for standard VTLN, the computation of the distance measure yields a relative distance measure of 0.25. As can be seen in Figure 6.11 this value is approximately obtained with a choice of $\nu = 0.25$ in these experiments. To allow for the generation of one-parameter warping functions for a comparison with the piecewise-linear warping function a PCA (see Section 2.3.1) is used for the extraction of the basis vectors on order of their corresponding eigenvalues. As an example, the three basis vectors with the largest eigenvalues for $\nu = 8$ are shown in Figure 6.12. As depicted in Figure 6.13 by using the first three basis vectors more than 99.9% of the energy is captured and already the use of only the first basis vector captures more than 99.4% of the original signal energy. Thus, the first PCA component $\boldsymbol{v}$ is used for generating a set of N warping functions $\mathcal{W}$,

$$\mathcal{W} = \{\, \beta_k \boldsymbol{v} \mid k = 1, 2, \ldots, N \,\}, \tag{6.39}$$

where β_k is chosen such that an appropriate warping range is covered. In these experiments a set of 17 warping functions was heuristically determined for each considered ν by choosing linearly spaced weights β_k such that the minimum and maximum average displacement of the resulting warping functions (according to Equation (6.39)) is -4.5 or $+4.5$, respectively. The resulting set of warping functions are shown in Figure 6.14. The impact of different choices for ν are

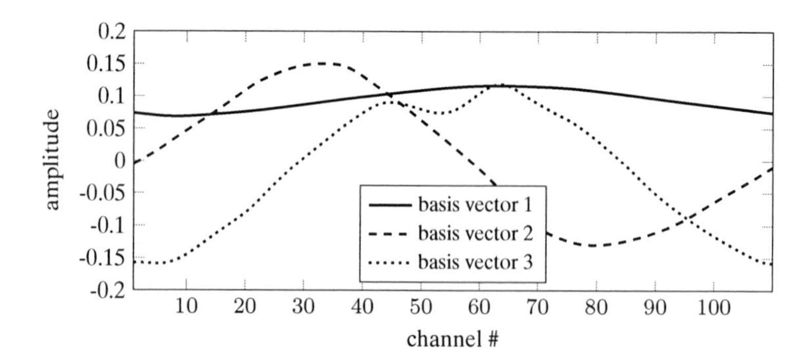

Figure 6.12.: The three basis vectors with the largest eigenvalues for $\nu = 8$.

clearly visible: The larger the regularizing parameter is chosen, the smoother are the resulting deformations. For a visual comparison, Figure 6.15 shows the piecewise-linear warping function for warping factors $\alpha_s = 0.88, 0.9, \ldots, 1.12$ as described in Section 2.5.1 mapped on an ERB scale. A difference between the model-based and the piecewise-linear warping functions is the channel number with maximum displacement: In case of the piecewise-linear warping function, the largest displacement occurs at channel indices in the range of 100 to 110 in with this setup. In contrast to that, the largest displacement, in case of the model-based warping function, occurs at channel indices in the range of 50 to 70. The model-based warping functions $\mathcal{W}$ are used for speaker-adaptive training (SAT) and for a maximum-likelihood grid-search VTLN (Welling et al., 2002) during the decoding stage.

Phoneme recognition experiments were conducted on the TIMIT corpus. Following the standard procedure, the SA sentences were not considered and the phoneme set was folded from 48 to 39 phonemes for accuracy evaluation. Three-state left-to-right monophone models with diagonal covariances and a bigram language model were used. Here, monophone models were used to allow for a higher computational efficiency. For the TF analysis, the procedure that was used for the analysis of the synthesized utterances was followed and a 110-channel gammatone filter bank was used. Twelve cepstral coefficients were computed from the filter bank output together with log-energy and Δ, $\Delta\Delta$-features.

The results of the phoneme recognition experiments are shown in Table 6.2. Baselines are shown in the upper two rows of the table and illustrate the benefits of VTLN. The remaining rows show accuracies for different choices of the regularizer weight ν. It can be seen that the highest accuracy for these experiments is obtained with a choice

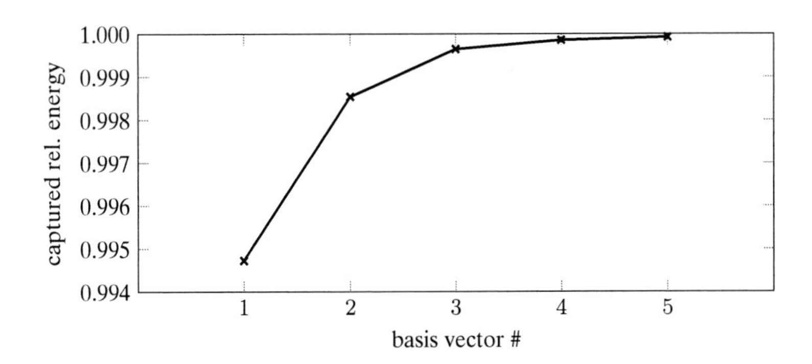

Figure 6.13.: The captured energy when using up to five basis vectors obtained with PCA for $\nu = 8$.

of $\nu = 8$. According to the MPSSWE significance test (see Section 2.4.4) this slight improvement in accuracy is significant.

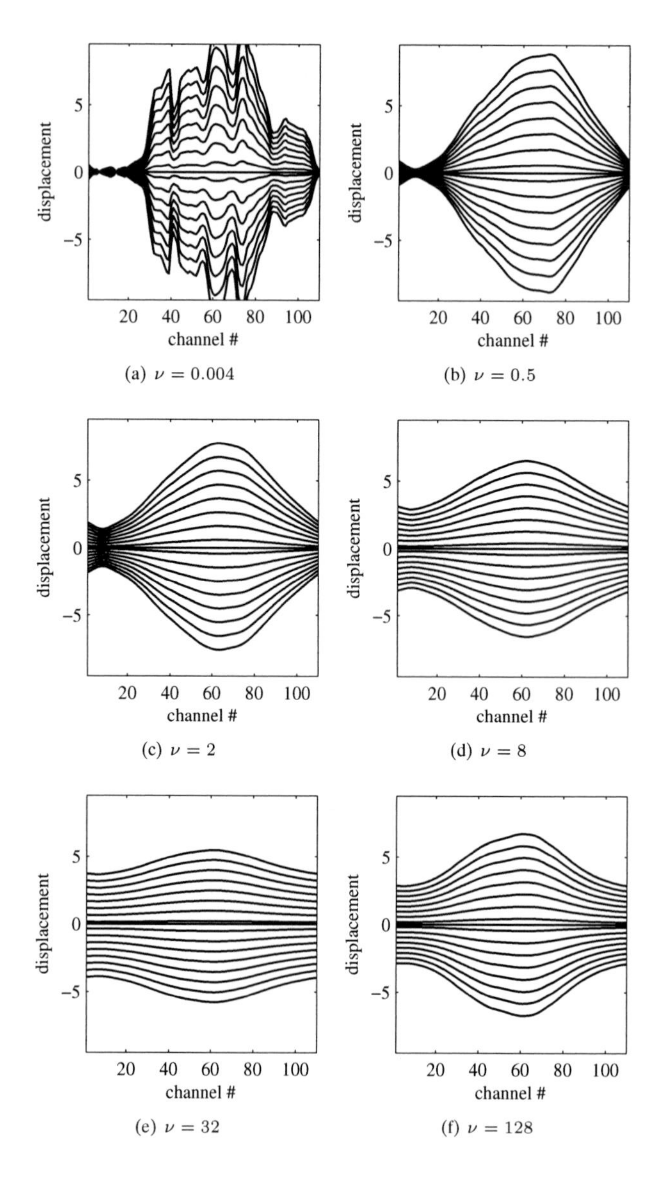

(a) $\nu = 0.004$ (b) $\nu = 0.5$

(c) $\nu = 2$ (d) $\nu = 8$

(e) $\nu = 32$ (f) $\nu = 128$

Figure 6.14.: Set of resulting warping functions for different choices of the regularizer weight ν.

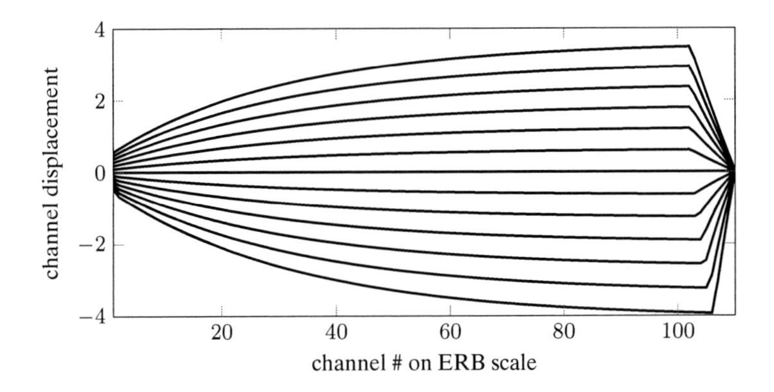

Figure 6.15.: Piecewise-linear (on a linear scale) warping function mapped to an ERB scale for warping factors $\alpha_s = 0.88, 0.9, \ldots, 1.12$.

Table 6.2.: Results of the phoneme recognition experiments.

warping function	**accuracy [%]**
-	66.4
(piecewise-)linear	68.0
elastic, $\nu = 0.004$	67.7
elastic, $\nu = 0.5$	68.0
elastic, $\nu = 2$	68.2
elastic, $\nu = 8$	**68.3**
elastic, $\nu = 32$	68.1
elastic, $\nu = 128$	68.0

7 Summary and Outlook

The performance of automatic speech recognition systems does not yet reach the human performance. A major cause for this is given by the huge amount of variability in the speech signals that occurs under real-world conditions. The different kinds of variabilities were described and grouped in Chapter 1 of this work. One of these variabilities is the length of the vocal tract, which has a major focus in this work. Until today, robustness against the effects of different VTLs in an ASR system is commonly achieved by either normalizing the features during or after their extraction or by adapting the models of the decoder. The principles of normalization and adaptation together with other basic methods of todays ASR systems were described in Chapter 2. The experimental setups for the considered recognition tasks in this work were presented in Chapter 3 together with baseline accuracies of common standard feature types for ASR. Furthermore, different auditory filter banks were evaluated with respect to the recognition performance of cepstral coefficients. The considered filter banks were the mel and the gammatone filter banks, as well as the static and the dynamic-compressive gammachirp filter bank. In these experiments, the gammatone filter bank lead to the highest accuracies, and, hence, was used by the feature extraction methods described in the rest of the following chapters. In the beginning of Chapter 4 an introduction into the principles of invariance transforms was given first, followed by an overview of related works in the field of invariant feature extraction methods for ASR. The assumption that different VTLs lead to translations along the subband-index axis of a time-frequency (TF) representation is fundamental for the presented invariant feature extraction methods. Because it is not clear which scale is best suited to this assumption, various scales including the log, mel, and ERB scale were evaluated in Chapter 4 with respect to the recognition rate of an ASR system with translational VTLN. The results showed that both the mel and the ERB scale yield similar accuracies, and both scales perform significantly better than the log scale. In addition, the use of gammatone filters lead to higher accuracies than the use of triangular ones, which supports the results of the experiments from Chapter 3.

7.1. Invariance Transforms for Feature Extraction in ASR

Three different translation-invariant methods for the extraction of features for speaker-independent ASR were described in Sections 4.4, 4.5, and 4.6, respectively.

The first method is based on transforms of the class $\mathbb{C}T$ (Müller and Mertins, 2010b). Translation-invariance is achieved with this method with the choice of two commutative operators and is based on the fast Walsh-Hadamard transform. Here, the use of the operators that are known from the rapid transform turned out to yield the highest accuracies. Especially under mismatching training-test conditions (with respect to the average vocal tract length) the CT-based features outperformed MFCC features clearly. These accuracies were further increased when the transform was applied on multiple scales of the time-frequency (TF) representation and when the feature vector was supplied by the correlation-based VTLI features from Rademacher et al. (2006).

The second method (Müller et al., 2009) is based on the so-called generalized cyclic transform (GCT). The proposed feature-extraction method consists of two steps for obtaining translation invariance: First, the spectral values of the current frame are linearly transformed by a generalized characteristics matrix (GCM). Second, a translation-invariant spectrum is computed from the transformed input signal. In the experimental part, different parametrizations for the GCM where evaluated, as well as two different ways from computing the translation-invariant spectrum. The multi-scale approach as used by the CT-based features assumes that translations occur along the whole frequency range of the TF representation. The "subframing scheme" used for the GCT features assumes that these translations occur only within certain intervals of the whole frequency range. The GCT-based extraction was performed on these intervals. The sole use of GCT-based features only leads to performance improvements in case of mismatching training-test conditions.

The combination with the correlation-based VTLI features showed improvements also for matching training-test conditions and outperformed MFCC features. The increase in discriminative information by the GCT-based features was also supported by the observation that the combination of GCT- and VTLI-features also outperforms the accuracies obtained with the VTLI-features only.

The first and second invariant extraction methods make use of invariance transforms that were specifically designed for translation-invariance. In contrast to that, the third feature extraction method, which was proposed in Chapter 4, makes use of a more general approach for computing invariant features. This approach is known as "invariant integration" and has a major focus in this work (Müller and Mertins, 2009a, 2010a, 2011a). The main idea of this method is to integrate (possibly nonlinear) functions of the spectral values from a certain temporal and spectral context. Because the proposed IIFs are based on monomials, their computation has a very low complexity. However, the definition of these features has a high degree of parametric freedom, so that an appropriate feature-selection method is needed. A method based on a linear classifier was used that iteratively enhances a feature set of a fixed size. The experiments showed that contextually selected MFCCs yield much higher accuracies in matching as well as in mismatching scenarios than standard MFCCs and LDA-reduced concatenated MFCCs. Invariant-integration features can be seen as a further refinement that tries to find important spectral cues within a certain temporal context and enhances the robustness of the resulting features to VTL changes. When no speaker-adaptation was used, the experiments showed a superior performance of the IIFs compared to cepstral coefficients (MFCCs and GTCCs) in matching, and especially in mismatching training-testing conditions. In the matching scenario, IIFs of order one perform better than IIFs based on higher order monomials. However, using higher order monomials yields better performances on mismatching training-testing scenarios. The experiments showed that the combination of IIFs with MLLR and/or VTLN further increases their accuracy. Within the experiments, the IIF-based systems lead to the highest accuracies and outperformed the cepstral coefficient-based systems in matching training-test conditions for the TIMIT phone-recognition task by more than one percentage point. The last part of the experiments showed that the TIMIT-based IIF sets can equally well be used for word recognition on the TIDIGITS corpus. Here, the IIF-based systems also yield accuracies that are higher than those of the MFCC-based system without and with VTLN.

A performance comparison of the different feature types on three different corpora was conducted at the end of Chapter 4. The results showed that the IIFs outperform the CT- and GCT-based features in the experiments without and with MLLR on TIMIT. The IIFs were also superior in the experiments on the OLLO task. System combination experiments with the ROVER method were also conducted at the end of Chapter 4 (Müller and Mertins, 2011c). The three presented invariant feature extraction methods from this work, as well as VTLI features were considered in combination with standard MFCC and PLP features. The back-ends of all systems were kept the same. It was shown that the combination of 1-best hypotheses of the systems with ROVER yields relative improvements in performance by up to eleven percent.

7.2. Noise Robustness of the Invariant Feature Types

Noise robustness of the feature extraction methods used by ASR systems is critical for many practical applications. In the beginning of Chapter 5 the presented feature extraction methods were first evaluated with respect to their performance under noisy conditions without any enhancement methods. It was shown that IIFs achieve the highest accuracies for SNRs down to 0 dB and perform better under all noise conditions than MFCC features. Also, it was shown that the addition of a subset of the VTLI features further increases the accuracies of the invariant feature types under all noise conditions. Afterwards, feature-space approaches that are commonly used for enhancing the noise robustness in ASR systems were evaluated. The effects of mean normalization (MN) and variance normalization (VN), as well as RASTA filtering, power-normalization (PN), and power-bias subtraction (PBS) were evaluated for the integration-based features. The results showed that the combination of PN, PBS, and MN leads to the highest accuracies among the considered enhancement methods. The achieved accuracies were also higher than for PNCC features.

The second part of Chapter 5 evaluated the use of a more sophisticated auditory model in combination with IIFs. The "auditory image model" (AIM) is a computational model of the human auditory processing pathway that represents a speech signal at every time instance within a two-dimensional space and is scale covariant. This space is called the "stabilized auditory image" (SAI). Motivated by other works that used the SAI as basis for extracting features for ASR, it was shown how the concept of invariant integration can be applied within the SAI space of this model. ASR experiments were conducted under different noise conditions on the Aurora-2 task. The results showed that the invariant features based on the AIM have a higher robustness against noise than MFCC, but perform worse than PNCC features. However, the combination of AIM-based integration features with power-normalized integration features yielded the highest accuracies under noisy conditions within the considered experiments in this work.

7.3. Estimation of the Spectral Effects due to Vocal-Tract Length Changes

A central assumption of the commonly used VTLN with a piecewise-linear warping function is the linear dependency between the resonance frequencies and the vocal tract length. This assumption stems from a model of the vocal tract as a lossless, uniform tube. Chapter 6 presented two different approaches that enhance the standard VTLN method without the assumption made by a lossless, uniform tube model.

The first approach was data-driven and makes use of elastic registration to relate a given spectral profile and a desired profile in terms of a displacement field with each other (Müller and Mertins, 2012a). The key idea of this method is to transform the TF representation of a given utterance such that it is more similar to a hypothesized optimal TF representation. Elastic registration is used to estimate a context-dependent, nonparametric transform for each frame. The hypothesis is generated with a standard VTLN method. Thus, the proposed "elastic VTLN" can be used as an additional enhancement stage within an ASR system. The results of the experiments showed that the proposed method is able to increase the accuracy of a monophone-based ASR system such that it reaches the performance of a system with triphone models and without VTLN. In an additional experiment, a set of global warping functions was computed from the registration based transforms. Instead of using the piecewise-linear warping function, these estimated warping functions were used in the standard grid-search VTLN. The results showed only slight improvements in accuracy compared to the standard VTLN approach. This indicates the importance of a context-dependent application of VTLN as done by the proposed elastic VTLN approach.

The second method within Chapter 6 (Müller and Mertins, 2012b) is based on a lossy model of the vocal tract, which was originally developed for articulatory speech synthesis. The model can simulate vocal tracts of different lengths and was used to generate a set of synthetic phones. As done by the data-driven approach, elastic registration was also used in this approach to describe the relation between the average spectral profiles of the same phonemes, but which were produced from vocal tracts with different lengths. The estimated transforms were used to estimate global warping functions, which, in turn, were used as replacement of the piecewise-linear warping function in the standard grid-search VTLN method. The results of this experiment showed slight improvements, which were comparable to those of experiments with the data-driven approach in which global warping functions were estimated.

7.4. Outlook

During the last decades, a lot of effort has been put into the enhancement of the various parts of the back-end. In contrast, the methods used in the front-ends of the vast majority of state-of-the-art ASR systems still makes use of methods that have been developed decades ago. Their underlying models have clear shortcomings with respect to their anatomical and psychoacoustic analogies. This work contributes to the field of more advanced feature extraction methods for ASR and gives possible directions for future work.

The use of complex-valued TF representations for the feature extraction is possible for all three described feature extraction methods: Transforms of the class $\mathbb{C}T$ can be applied on complex input signals. The same holds for the GCT- and invariant-integration based features. By using appropriately chosen characteristic coefficients, the GCT-based extraction method could apply a complex transformation matrix prior to the computation of the average or extended group spectrum. The invariant-integration approach used by the presented IIFs allows for the integration over arbitrary transformation groups. In this work, the group of translations was considered only and increases in performance were already observed. Instead of translations, other transformations that better model the spectral effects due to different VTLs could be used for this integration approach. A critical property of all three presented feature extraction methods is the high dimensionality of the resulting feature vectors. For the experiments in this work, an LDA was used to reduce the dimensionality of the feature vectors. This procedure was a compromise between computational efficiency and recognition accuracy. With respect to the ASR research community, there are works that make use of various other dimensionality reduction methods (for example, Kumar and Andreou, 1998; Saon et al., 2000; Heckmann and Gläser, 2011) and might yield improvements in performance. Also, information theoretic transforms (Torkkola, 2003; Ozertem et al., 2006; II et al., 2006) have shown promising advances in contrast to the traditional feature transforms like PCA or LDA. Their application for selecting invariant features or for dimensionality reduction seems promising in the context of this work. Another stage of all feature extraction methods for ASR is the computation of a TF representation. The mel and gammatone filter banks, which are used in the standard feature extraction methods nowadays, do not explain as many psychoacoustic findings as state-of-the-art auditory models like the dynamic-compressive gammachirp filter bank or the auditory image model. Nevertheless, it was shown in this, as well as in other works (for example, Walters, 2011) that the use of these more sophisticated models does not necessarily yield a better performance for ASR systems. It would be interesting to analyze the reasons for this discrepancy in future works. This might allow one to take further advantages of these auditory models. For example, the AIM-based IIFs, which were described in this work, use a gammatone filter bank for the simulation of the basilar membrane motion. An optimization of the parameters of the more sophisticated dynamic-compressive gammachirp filter bank could improve the overall performance of the AIM-based IIFs.

Noise robustness of the presented feature extraction methods was evaluated within this work with additively distorted utterances. Future work in the field of noise robustness for invariant feature extraction methods could consider tasks in which the Lombard effect is observable. Especially the IIFs have a high flexibility due to their general parametrization. One possibility to take advantage of this fact would be to make the parametrization adaptive such that the robustness against certain other variabilities than the VTL is increased.

The proposed "elastic VTLN" as an enhancement to standard VTLN is a data-driven approach for estimating the spectral effects of different VTLNs. It has lead to large performance improvements in the experimental part. The choice of both the distance measure as well as the regularization method within this approach can have a considerable effect on the solution. It is shown in the experimental part that the choices for this work yield promising results. However, additional experiments will have to show if other measures or regularizers are even more beneficial. Another common approach for registration methods is the introduction of an additional penalty term. The objective function used here does not account for energy preservation (with respect to the spectral values) during the computation of the transformation. An appropriate penalty term could take care for this. Due to the normalization during the subsequent feature extraction in this work, one can assume that the effect of not considering energy preservation is mitigated. Nevertheless, a subtle analysis and refinement of the method might provide further performance improvements. The use of a larger corpus would allow for the training of reference triphone models instead of monophone models. Also, a two-pass VTLN approach could be used and might yield another improvement in accuracy.

The data-driven enhancement method for VTLN could be further developed in several aspects: With articulatory models depending on the results of MRI studies for determining the area function parameters, measurements from additional MRI studies might yield improved model parameters with better generalization capabilities. In its current implementation, different VTLs are simulated by choosing different numbers of tube elements and then linearly interpolating the diameter function $\Omega(k)$ and basis functions $\Phi_1(k), \Phi_2(k)$. However, it was shown, for example by Boë et al. (2006), that the relation of the pharyngeal cavity length to oral cavity length is different for women and men. Thus, a modeling of different VTLs that accompanies for these effects would yield a more realistic simulation. There exist also other articulatory speech synthesis models that could be considered for a model-driven method as described in this section. One particular model that has proven to produce naturally sounding utterances is described in the work of Birkholz (2005). Nevertheless, also the model of Birkholz relies on the results of MRI studies.

As mentioned for the data-driven enhancement method above, the distance measure as well as the regularizing term of the registration method might be modified in order to improve the estimation. As was observed in the experiments described in Section 6.1.3, a nonglobal application of warping functions for VTLN is critical for increasing the performance of the ASR system. Thus, a model-driven, phoneme-dependent estimation and application of warping functions seems promising for increasing the normalization capabilities of the proposed method.

Bibliography

(Apr. 2010). *ASCII*. Oxford Dictionaries.

(Oct. 2010). *Speech Recognition Scoring Toolkit (SCTK) Version 2.4.0*. National Institute of Standards and Technology (NIST).

(Nov. 2011). *CMUdict: Carnegie Mellon University Pronouncing Dictionary*. Carnegie Mellon University.

(Oct. 2011). *Matched Pairs Sentence-Segment*. National Institute of Standards and Technology (NIST).

A. Acero and R. Stern (Apr. 1991). Robust speech recognition by normalization of the acoustic space. In *Int. Conf. Acoustics, Speech, and Signal Processing*, volume 2, pages 893 –896. Toronto, Canada.

N. Ahmed, K. R. Rao, and A. L. Abdussattar (Sept. 1971). BIFORE or Hadamard transform. *IEEE Trans. Audio Electroacoustics*, 19:225–234.

N. Ahmed, K. R. Rao, and A. L. Abdussattar (Aug. 1973). On cyclic autocorrelation and the Walsh-Hadamard transform. *IEEE Trans. Electromagnetic Compatibility*, EMC-15(3):141–146.

N. U. Ahmed and K. R. Rao (1975). *Orthogonal Transforms for Digital Signal Processing*. Springer, New York, USA.

F. Alleva, X. Huang, M. yuh Hwang, and L. Jiang (1998). Can continuous speech recognizers handle isolated speech? *Speech Communication*, 26:183–189.

T. Anastasakos, J. McDonough, R. Schwartz, and J. Makhoul (Oct. 1996). A compact model for speaker-adaptive training. In *Proc. Conf. Spoken Language (ICSLP 96)*, volume 2, pages 1137–1140.

L. R. Bahl, P. V. de Soutza, P. S. Gopalakrishnan, D. Nahamoo, and M. A. Picheny (1991). Context dependent modeling of phones in continuous speech using decision trees. In *Proc. Workshop Speech and Natural Language*, HLT '91, pages 264–269. Association for Computational Linguistics, Stroudsburg, USA.

L. E. Baum, T. Petrie, G. Soules, and N. Weiss (Feb. 1970). A maximization technique occurring in the statistical analysis of probabilistic functions of Markov chains. *The Annals of Mathematical Statistics*, 41(1):164–171.

J. Benesty, M. M. Sondhi, and Y. Huang, editors (2008). *Handbook of Speech Processing*. Springer, Berlin, Germany.

M. Benzeghiba, R. D. Mori, O. Deroo, S. Dupont, T. Erbes, D. Jouvet, L. Fissore, P. Laface, A. Mertins, C. Ris, R. Rose, V. Tyagi, and C. Wellekens (Oct.-Nov. 2007). Automatic speech recognition and speech variability: a review. *Speech Communication*, 49(10-11):763–786.

P. Birkholz (2005). *3-D Artikulatorische Sprachsynthese*. Ph.D. thesis, Fakultät für Informatik und Elektrotechnik, Universität Rostock, Rostock, Germany.

P. Birkholz and D. Jackèl (2006). Noise sources and area functions for the synthesis of fricative consonants. *Rostocker Informatik Berichte*, 30:17 – 23.

P. Birkholz, D. Jackèl, and B. Kröger (May 2007). Simulation of losses due to turbulence in the time-varying vocal system. *IEEE Trans. Audio, Speech, and Language Processing*, 15(4):1218–1226.

C. Bishop (2006). *Pattern Recognition and Machine Learning*. Information Science and Statistics. Springer, New York, USA.

S. Bleeck, T. Ives, and R. D. Patterson (2004). Aim-mat: the auditory image model in MATLAB. *Acta Acustica United with Acustica*, 90:781–788.

L.-J. Boë, J. Granat, P. Badin, D. Autesserre, D. Pochic, N. Zga, N. Henrich, and L. Ménard (Dec. 2006). Skull and vocal tract growth from newborn to adult. In *Proc. 7th Int. Seminar on Speech Production (ISSP7)*, pages 75–82. Ubatuba, Brazil.

L. Breiman, J. Friedman, C. J. Stone, and R. Olshen (1984). *Classification and Regression Trees*. The Wadsworth statistics/probability series. Chapman and Hall, Boca Raton, USA.

M. Brookes (Nov. 2011). *VOICEBOX: Speech Processing Toolbox for MATLAB*. Department of Electrical & Electronic Engineering, Imperial College, London, UK.

P. F. Brown, P. V. deSouza, R. L. Mercer, V. J. D. Pietra, and J. C. Lai (Dec. 1992). Class-based N-gram models of natural language. *Comput. Linguist.*, 18:467–479.

H. Burkhardt and X. Müller (Oct. 1980). On invariant sets of a certain class of fast translation-invariant transforms. *IEEE Trans. Acoustic, Speech, and Signal Processing*, 28(5):517–523.

H. Burkhardt and S. Siggelkow (2001). *Nonlinear Model-Based Image/Video Processing and Analysis*, chapter Invariant features in pattern recognition – fundamentals and applications, pages 269–307. John Wiley & Sons.

S. Chen and J. Goodman (1999). An empirical study of smoothing techniques for language modeling. *Computer Speech & Language*, 13(4):359–393.

L. Cohen (December 1993). The scale representation. *IEEE Trans. Signal Processing*, 41(12):3275–3292.

P. J. Davis (1979). *Circulant matrices*. Chelsea Publishing, New York, USA.

S. Davis and P. Mermelstein (Aug. 1980). Comparison of parametric representations for monosyllabic word recognition in continuously spoken sentences. *IEEE Trans. Acoustics, Speech and Signal Processing*, 28(4):357–366.

J. R. Deller, J. G. Proakis, and J. H. L. Hansen (1993). *Discrete-Time Processing of Speech Signals*. Macmillan, New York.

L. Deng, A. Acero, M. Plumpe, and X. Huang (Oct. 2000). Large-vocabulary speech recognition under adverse acoustic environments. In *Proc. Int. Conf. Spoken Language Processing (Interspeech-2000)*, pages 806–809. Beijing, China.

J. Droppo, A. Acero, and L. Deng (2001). Evaluation of the SPLICE algorithm on the Aurora-2 database. In *Proc. Eurospeech Conference*.

J. Droppo, L. Deng, and A. Acero (May 2002). Uncertainty decoding with SPLICE for noise robust speech recognition. In *Proc. Int. Conf. Acoustics, Speech, and Signal Processing*, pages I–57 – I–60. Orlando, USA.

R. O. Duda, P. E. Hart, and D. G. Stork (2001). *Pattern Classification*. John Wiley & Sons, 2nd edition.

E. Eide and H. Gish (May 1996). A parametric approach to vocal tract length normalization. In *Proc. Int. Conf. Acoustics, Speech, and Signal Processing*, volume 1, pages 346–349. Atlanta, GA.

D. P. W. Ellis (Nov. 2005). *PLP and RASTA (and MFCC, and inversion) in Matlab using melfcc.m and invmelfcc.m*.

D. P. W. Ellis (Jun. 2009). Gammatone-like spectrograms. web resource: http://www.ee.columbia.edu/~dpwe/resources/matlab/gammatonegram.

G. Evermann and P. C. Woodland (2000). Posterior probability decoding, confidence estimation and system combination. In *Proc. Speech Transcription Workshop*. College Park, USA.

M. Fang and G. Häusler (Mar. 1989). Modified rapid transform. *Applied Optics*, 28(6):1257–1262.

G. Fant (1973). *Speech sounds and features.* MIT Press, Cambridge, USA.

G. Fant, J. Liljencrants, and Q.-g. Lin (1985). A four-parameter model of glottal flow. *STL-QPSR*, 4(1985):1–13.

J. Feng, B. Ramabhadran, J. Hansen, and J. Williams (Jan. 2012). Trends in speech and language processing [in the spotlight]. *Signal Processing Magazine, IEEE*, 29(1):177–179.

J. G. Fiscus (Dec. 1997). A post-processing system to yield reduced word error rates: Recognizer output voting error reduction (ROVER). In *Proc. IEEE Int. Workshop Automatic Speech Recognition and Understanding (ASRU)*, pages 347–354. Santa Barbara, USA.

H. Fletcher (1953). *Speech and Hearing in Communication.* Van Nostrand Princeton, New Jersey.

K. Fukunaga (2001). *Introduction to Statistical Pattern Recognition.* Computer Science and Scientific Computing. Academic Press, San Diego, USA, 2nd edition.

M. Gales and S. Young (2008). *The application of hidden Markov models in speech recognition*, volume 1. Now Publishers Inc., Hanover, USA.

M. J. F. Gales (Sept. 1995). *Model-Based Techniques for Noise Robust Speech Recognition.* Ph.D. thesis, Cambridge University, Cambridge.

M. J. F. Gales (Apr. 1998). Maximum likelihood linear transformations for HMM-based speech recognition. *Computer Speech and Language*, 12(2):75–98.

M. J. F. Gales, X. Liu, R. Sinha, P. C. Woodland, K. Yu, S. Matsoukas, T. Ng, K. Nguyen, L. Nguyen, J.-L. Gauvainy, L. Lamely, and A. Messaoudi (Apr. 2007). Speech recognition system combination for machine translation. In *Proc. Int. Conf. Audio, Speech, and Signal Processing*. Honolulu, Hawaii.

J. Gamec and J. Turan (Dec. 1996). Use of Invertible Rapid Transform in Motion Analysis. *Radioengineering*, 5(4):pp. 21–27.

Y. Gao, M. Padmanabhan, and M. Picheny (Sept. 1997). Speaker adaptation based on pre-clustering training speakers. In *Proc. European Conf. Speech Communicatoin and technology (EUROSPEECH-1997)*, pages 2091–2094. Rhodes, Greece.

J. S. Garofolo, L. F. Lamel, W. M. Fisher, J. G. Fiscus, D. S. Pallett, and N. L. Dahlgren (1993). *DARPA TIMIT acoustic phonetic speech corpus.* Linguistic Data Consortium, Philadelphia.

L. Gillick and S. Cox (May 1989). Some statistical issues in the comparison of speech recognition algorithms. In *Int. Conf. Acoustics, Speech, and Signal Processing*, volume 1, pages 532 -535.

B. R. Glasberg and B. C. J. Moore (1990). Derivation of auditory filter shapes from notched-noise data. *Hearing Research*, 47:103–138.

G. H. Golub and C. F. Van Loan (1996). *Matrix Computations*. Johns Hopkins University Press, Baltimore, USA, 3rd edition.

R. A. Gopinath (May 1998). Maximum likelihood modeling with gaussian distributions for classification. In *Proc. Int. Conf. Acoustics, Speech, and Signal Processing*, pages 661–664. Seattle, USA.

T. Gramss (Oct. 1991). Word recognition with the feature finding neural network (FFNN). In *Proc. IEEE Workshop Neural Networks for Signal Processing*, pages 289–298. Princeton, USA.

R. Haeb-Umbach and H. Ney (Mar. 1992). Linear discriminant analysis for improved large vocabulary continuous speech recognition. In *Proc. Int. Conf. Acoustics, Speech, and Signal Processing*, volume 1, pages 13–16. San Francisco, USA.

T. Hain, P. C. Woodland, T. R. Niesler, and E. W. D. Whittaker (May 1999). The 1998 HTK system for transcription of conversational telephone speech. In *Proc. Int. Conf. Audio, Speech, and Signal Processing*, pages 57–60. Phoenix, USA.

A. K. Halberstadt (Nov. 1998). *Heterogeneous Acoustic Measurements and Multiple Classifiers for Speech Recognition*. Ph.D. thesis, Massachusetts Institute of Technology.

A. K. Halberstadt and J. R. Glass (1998). Heterogeneous measurements and multiple classifiers for speech recognition. In *Fifth Int. Conf. Spoken Language Processing*. ISCA.

J. H. L. Hansen (Nov. 1996). Analysis and compensation of speech under stress and noise for environmental robustness in speech recognition. *Speech Communications, Special Issue on Speech Under Stress*, 20(2):151–170.

F. Harris (1978). On the use of windows for harmonic analysis with the discrete fourier transform. *Proc. IEEE*, 66(1):51–83.

H. O. Hartley (Jun. 1958). Maximum likelihood estimation from incomplete data. *Biometrics*, 14(2):174–194.

M. Heckmann and C. Gläser (Sept. 2011). Discriminant sub-space projection of spectro-temporal speech features based on maximizing mutual information. In *Proc. Interspeech*, pages 225–228. Florence, Italy.

G. E. Henter and W. B. Kleijn (Sept. 2011). Intermediate-state HMMs to capture continuously-changing signal features. In *Proc. Interspeech*, pages 1817–1820. Florence, Italy.

H. Hermansky (Apr. 1990). Perceptual linear predictive (PLP) analysis of speech. *J. Acoustical Society of America*, 87(4):1738–1752.

H. Hermansky and N. Morgan (Oct. 1994). Rasta processing of speech. *IEEE Trans. Speech and Audio Processing*, 2(4):578–589.

H. Hermansky, N. Morgan, A. Bayya, and P. Kohn (Mar. 1992). Rasta-plp speech analysis technique. *Proc. Int. Conf. Acoustics, Speech, and Signal Processing*, 1:121–124.

H.-G. Hirsch and D. Pearce (2000). The AURORA experimental framework for the performance evaluation of speech recognition systems under noisy conditions. In *ISCA Tutorial and Research Workshop ASR2000: Challenges for the new Millenium*, pages 181–188.

B. Hoffmeister, T. Klein, R. Schlüter, and H. Ney (Sept. 2006). Frame based system combination and a comparison with weighted rover and cnc. In *Proc. Int. Conf. Spoken Language Processing (Interspeech-2006)*. Pittsburgh, USA.

C. Huang, T. Chen, S. Li, E. Chang, and J. Zhou (Sept. 2001a). Analysis of speaker variability. In *Proc. EUROSPEECH-2001*, pages 1377–1380. Aalborg, Denmark.

X. Huang, A. Acero, and H. Hon (2001b). *Spoken Language Processing: A Guide to Theory, Algorithm, and System Development.* Prentice Hall PTR Upper Saddle River, New York, USA.

A. Hurwitz (1897). Ueber die Erzeugung der Invarianten durch Integration. *Nachrichten von der Königl. Gesellschaft der Wissenschaften zu Göttingen, Mathematisch-physikalische Klasse*, pages 71–90.

K. E. H. II, D. Erdogmus, K. Torkkola, and J. C. Principe (Sept. 2006). Feature extraction using information-theoretic learning. *IEEE Transactions on Pattern Analysis and Machine Intelligence*, 28(9):1385–1392.

T. Irino (May 1996). A 'gammachirp' function as an optimal auditory filter with the mellin transform. *Int. Conf. Acoustics, Speech, and Signal Processing*, 2:981–984.

T. Irino and R. Patterson (March 2002). Segregating information about the size and the shape of the vocal tract using a time-domain auditory model: The stabilised wavelet-Mellin transform. *Speech Communication*, 36(3):181–203.

T. Irino and R. D. Patterson (1997). A time-domain, level-dependent auditory filter: The gammachirp. *J. Acoustical Society of America*, 101(1):412–419.

T. Irino and R. D. Patterson (Aug. 1999). Extracting size and shape information of sound source in an optimal auditory processing model. In *Workshop on Computational Auditory Scene Analysis (CASA), Int. Joint Conf. Artificial Intelligence (IJCAI'99)*. Stockholm, Sweden.

T. Irino and R. D. Patterson (Nov. 2006). A dynamic compressive gammachirp auditory filterbank. *IEEE Trans. Audio, Speech and Language Processing*, 14(6):2222–2232.

ITU (Nov. 2001). ITU recommendation G.712, transmission performance characteristics of pulse code modulation channels.

S. M. Katz (Mar. 1987). Estimation of probabilities from sparse data for the language model component of a speech recognizer. *IEEE Trans. Acoust., Speech, Signal Processing*, 35:400–401.

J. Kelly and C. Lochbaum (1963). Speech synthesis. In *Proc. Speech Communications Seminar*. Royal Institute of Technology, Stockholm, Sweden.

C. Kim (Dec. 2010). *Signal Processing for Robust Speech Recognition Motivated by Auditory Processing*. Ph.D. thesis, Language Technologies Institute, School of Computer Science, Carnegie Mellon University, Pittsburgh, USA.

C. Kim and R. M. Stern (Sept. 2009). Feature extraction for robust speech recognition using a power-law nonlinearity and power-bias subtraction. In *Proc. Interspeech 2009*, pages 28–31. Brighton, UK.

C. Kim and R. M. Stern (Mar. 2010a). Feature extraction for robust speech recognition based on maximizing the sharpness of the power distribution and on power flooring. In *Proc. Int. Conf. Audio, Speech, and Signal Processing*, pages 4574–4577. Dallas, USA.

C. Kim and R. M. Stern (Sep. 2010b). Open source matlab code for PNCC. web resource: http://www.cs.cmu.edu/~robust/archive/algorithms/ PNCC_ICASSP2010.

D. Y. Kim, S. Umesh, M. J. F. Gales, T. Hain, and P.Woodland (2004). Using VTLN for broadcast news transcription. In *Proc. Int. Conf. Speech and Language Processing*. Jeju, Koea.

M. Kleinschmidt (Sep. 2002). *Robust speech recognition based on spectro-temporal processing*. Ph.D. thesis, Universität Oldenburg.

M. Kleinschmidt and D. Gelbart (Sept. 2002). Improving word accuracy with Gabor feature extraction. In *Proc. Int. Conf. Spoken Language Processing*, pages 25–28.

B. J. Kröger, R. Winkler, C. Mooshammer, and B. Pompino-Marschall (May 2000). Estimation of vocal tract area function from magnetic resonance imaging: preliminary results. In *Proc. 5th Seminar on Speech Production: Models and Data*, pages 333–336. Seeon, Germany.

N. Kumar (1997). *Investigation of silicon auditory models and generalization of linear discriminant analysis for improved speech recognition*. Ph.D. thesis, The Johns Hopkins University, Baltimore, Maryland.

N. Kumar and A. G. Andreou (1998). Heteroscedastic discriminant analysis and reduced rank hmms for improved speech recognition. *Speech Commun.*, 26(4):283–297. ISSN 0167-6393.

P. Ladefoged (2001). *Vowels and Consonants: An Introduction to the Sounds of Languages*. Blackwell Publishers, Malden, USA.

A. D. Lawson, D. M. Harris, and J. J. Grieco (Sept. 2003). Effect of foreign accent on speech recognition in the nato n-4 corpus. In *Proc. EUROSPEECH-2003*, pages 1505–1508. Geneva, Switzerland.

K. F. Lee and H. W. Hon (Nov. 1989). Speaker-independent phone recognition using hidden Markov models. *IEEE Trans. Acoustics, Speech and Signal Processing*, 37(11):1641–1648.

K.-F. Lee, H.-W. Hon, and M.-Y. Hwang (1989). Recent progress in the SPHINX speech recognition system. In *Proc. workshop Speech and Natural Language*, HLT '89, pages 125–130. Association for Computational Linguistics, Stroudsburg, USA.

L. Lee and R. Rose (May 1996). Speaker normalization using efficient frequency warping procedures. In *Proc. Int. Conf. Audio, Signal, and Speech Processing*, volume 1, pages 353–356. Atlanta, USA.

L. Lee and R. C. Rose (Jan. 1998). A frequency warping approach to speaker normalization. *IEEE Trans. Speech and Audio Processing*, 6(1):49–60.

C. Leggetter and P. Woodland (Apr. 1995). Maximum likelihood linear regression for speaker adaptation of continuous density hidden Markov models. *Computer Speech and Language*, 9(2):171–185.

R. G. Leonard (1984). A database for speaker-independent digit recognition. In *Proc. Int. Conf. Audio, Speech, and Signal Processing 1984*, volume 3, pages 42.11.1–42.11.4. San Diego, USA.

V. Levenshtein (Feb. 1966). Binary codes capable of correcting deletions, insertions, and reversals. *Soviet Physics Doklady, Cybernetics and Control Theory*, 10(8):707–710.

J. Liljencrants (1985). *Speech synthesis with a reflection-type line analog*. Ph.D. thesis, Royal Institute of Technology, Stockholm, Sweden.

R. P. Lippmann (1997). Speech recognition by machines and humans. *Speech Communication*, 22(1):1–15.

V. Lohweg (2003). *Ein Beitrag zur effektiven Implementierung adaptiver Spektraltransformationen in applikationsspezifische integrierte Schaltkreise*. Ph.D. thesis, Technische Universität Chemnitz.

V. Lohweg and D. Müller (Jun. 2001). Nonlinear generalized circular transforms for signal processing and pattern recognition. In *Proc. IEEE Workshop Nonlinear Signal and Image Processing*. Baltimore.

V. Lohweg and D. Müller (Jan. 2002). A complete set of translation invariants based on the cyclic correlation property of the generalized circular transforms. In *Proc. 6th Digital Image Computing Techniques and Applications (DICTA'02)*, pages 134–138. Australian Pattern Recognition Society, Melbourne, Australia.

V. Lohweg, C. Diederichs, and D. Müller (Jan. 2004). Algorithms for hardware-based pattern recognition. *EURASIP J. Applied Signal Processing*, 2004:1912–1920.

E. Lombard (1911). Le signe de l'élévation de la voix. *Ann. Maladies Oreille, Larynx, Nez, Pharynx*, 37:101–119.

R. F. Lyon, M. Rehn, S. Bengio, T. C. Walters, and G. Chechik (2010). Sound retrieval and ranking using sparse auditory representations. *Neural Computation*, 22:2390–2416.

J. Makhoul (Jun. 1975). Spectral linear prediction: Properties and applications. *IEEE Trans. Acoust., Speech, Signal Processing*, 23:283–296.

L. L. Mangu (Apr. 2000). *Finding Consensus in Speech Recognition*. Ph.D. thesis, Johns Hopkins University, Baltimore, USA.

M. G. Maragakis and A. Potamianos (Sept. 2008). Region-based vocal tract length normalization for ASR. In *Proc. Interspeech-2008*, pages 1365–1368. Brisbane, Australia.

S. Martin, J. Liermann, and H. Ney (Sept. 1995). Algorithms for bigram and trigram word clustering. In *Proc. Eurospeech*, volume 2, pages 1253–1256. Madrid, Spain.

S. Mathur, B. Story, and J. Rodriguez (Sept. 2006). Vocal-tract modeling: Fractional elongation of segment lengths in a waveguide model with half-sample delays. *IEEE Trans. Audio, Speech, and Language Processing*, 14(5):1754 – 1762.

A. Mertins (1994). *Signal Analysis - Wavelets, Filter Banks, Time-Frequency Transforms and Applications*. John Wiley & Sons, Chichester, England.

A. Mertins and J. Rademacher (Nov. 27 -Dec. 1 2005). Vocal tract length invariant features for automatic speech recognition. In *Proc. 2005 IEEE Automatic Speech Recognition and Understanding Workshop*, pages 308–312. San Juan, Puerto Rico.

A. Mertins and J. Rademacher (May 2006). Frequency-warping invariant features for automatic speech recognition. In *Proc. IEEE Int. Conf. Acoustics, Speech, and Signal Processing*, volume V, pages 1025–1028. Toulouse, France.

N. Mesgarani, S. Shamma, and M. Slaney (May 2004). Speech discrimination based on multiscale spectro-temporal modulations. *Proc. Int. Conf. Acoustics, Speech, and Signal Processing*, 1:I–601–I–604.

B. T. Meyer, T. Jurgens, T. Wesker, T. Brand, and B. Kollmeier (2010). Human phoneme recognition depending on speech-intrinsic variability. *J. Acoustical Society of America*, 128(5):3126–3141.

A. Miguel, E. Lleida, R. Rose, L. Buera, and A. Ortega (Sept. 2005). Augmented state space acoustic decoding for modeling local variability in speech. In *Proc. Interspeech-2005*, pages 3009–2012. Lisbon, Portugal.

J. Modersitzki (2004). *Numerical Methods for Image Registration*. Oxford University Press, New York.

M. Mohri, F. Pereira, and M. Riley (2002). Weighted finite-state transducers in speech recognition. *Computer Speech & Language*, 16(1):69–88.

J. J. Monaghan, C. Feldbauer, T. C. Walters, and R. D. Patterson (Jul. 2008). Low-dimensional, auditory feature vectors that improve vocal-tract-length normalization in automatic speech recognition. *J. Acoustical Society of America*, 123(5):3066–3066.

B. C. J. Moore (1995). *Hearing*. Elsevier Academic Press, 2nd edition.

B. C. J. Moore and B. R. Glasberg (Mar. 1996). A revision of Zwicker's loudness model. *Acta Acustica united with Acustica*, 82(11):335–245.

P. J. Moreno, B. Raj, and R. M. Stern (May 1996). A vector taylor series approach for environment independent speech recognition. In *Proc. Int. Conf. Audio, Speech, and Signal Processing*, volume 2, pages 733–736. Atlanta, USA.

N. Morgan and H. Hermansky (Nov. 1992). RASTA extensions: Robustness to additive and convolutional noise. In *Proc. SPAC-1992*, pages 115–118. Cannes-Mandelieu, France.

F. Müller and A. Mertins (Sept. 2009a). Invariant-integration method for robust feature extraction in speaker-independent speech recognition. In *Proc. Int. Conf. Spoken Language Processing (Interspeech 2009-ICSLP)*, pages 2975–2978. Brighton, UK.

F. Müller and A. Mertins (Jun. 2009b). Nonlinear translation-invariant transformations for speaker-independent speech recognition. In *ISCA Tutorial and workshop on non-linear speech processing (NOLISP'09)*. Vic, Spain.

F. Müller and A. Mertins (Sept. 2010a). Invariant integration features combined with speaker-adaptation methods. In *Proc. Int. Conf. Spoken Language Processing (Interspeech 2010)*, pages 2622–2625. Makuhari, Japan.

F. Müller and A. Mertins (Feb. 2010b). Robust features for speaker-independent speech recognition based on a certain class of translation-invariant transformations. In J. Sole-Casals and V. Zaiats, editors, *Advances in Nonlinear Speech Processing*, volume 5933 of *LNAI*, pages 111–119. Springer, Heidelberg, Germany.

F. Müller and A. Mertins (2011a). Contextual invariant-integration features for improved speaker-independent speech recognition. *Speech Communication*, 53(6):830 – 841.

F. Müller and A. Mertins (Aug. 2011b). Noise robust speaker-independent speech recognition with invariant-integration features using power-bias subtraction. In *Proc. Int. Conf. Spoken Language Processing (Interspeech 2011-ICSLP)*, pages 1677–1680. Florence, Italy.

F. Müller and A. Mertins (Sept. 2011c). Robust continuous speech recognition through combination of invariant-feature based systems. In *Proc. German Conf. Speech Signal Processing (ESSV 2011)*, pages 229–236. Aachen, Germany.

F. Müller and A. Mertins (Sept. 2012a). Enhancing vocal tract length normalization with elastic registration for automatic speech recognition. In *Proc. Interspeech-2012*. Portland, USA.

F. Müller and A. Mertins (Mar. 2012b). On the use of a wave-reflection model for the estimation of spectral effects due to vocal tract length changes with application to automatic speech recognition. In *Proc. 38th German Annual Conf. Acoustics (DAGA 2012)*. Darmstadt, Germany.

F. Müller and A. Mertins (Mar. 2012c). On using the auditory image model and invariant-integration for noise robust automatic speech recognition. In *Proc. Int. Conf. Audio, Speech, and Signal Processing*, pages 4905–4908. Kyoto, Japan.

F. Müller, E. Belilovsky, and A. Mertins (Dec. 2009). Generalized cyclic transformations in speaker-independent speech recognition. In *Proc. IEEE Automatic Speech Recognition and Understanding Workshop*, pages 211–215. Merano, Italy.

E. Noether (Mar. 1915). Der Endlichkeitssatz der Invarianten endlicher Gruppen. *Mathematische Annalen*, 77(1):89–92.

A. V. Oppenheim and R. W. Schafer (1999). *Discrete-Time Signal Processing*. Prentice Hall, New Jersey, USA, 2nd. edition.

D. O'Shaughnessy (Nov. 1999). *Speech Communications: Human and Machine*. Wiley-IEEE Press, 2nd edition.

U. Ozertem, D. Erdogmus, and R. Jenssen (2006). Spectral feature projections that maximize shannon mutual information with class labels. *Pattern Recognition*, 39(7):1241–1252.

D. Pallet, W. Fisher, and J. Fiscus (Apr. 1990). Tools for the analysis of benchmark speech recognition tests. In *Int. Conf. Acoustics, Speech, and Signal Processing*, volume 1, pages 97 –100.

R. Patterson, R. van Dinther, and T. Irino (Sept. 2007). The robustness of bio-acoustic communication and the role of normalization. In *Proc. Int. Congress on Acoustics*, pages ppa–07–011. Madrid.

R. D. Patterson (2000). Auditory images: How complex sounds are represented in the auditory system. *J. Acoustical Society of Japan (E)*, 21(4):183–190.

R. D. Patterson, J. Nimmo-Smith, J. Holdsworth, and P. Rice (December 14-15 1987). An efficient auditory filterbank based on the gammatone function. In *Proc. Meeting of the IOC Speech Group on Auditory Modelling at RSRE*.

R. D. Patterson, K. Robinson, J. Holdsworth, D. McKeown, C. Zhang, and M. Allerhand (1992). Complex sounds and auditory images. In Y. Cazals, L. Demany, and K. Horner, editors, *Auditory Physiology and Perception. Advanced Bioscience*, volume 83, pages 429–446. Pergamon, Oxford.

R. D. Patterson, M. Unoki, and T. Irino (2003). Extending the domain of center frequencies for the compressive gammachirp auditory filter. *J. Acoustical Society of America*, 114(3):1529–1542.

J. Pender and D. Covey (Aug. 1992). New square wave transform for digital signal processing. *IEEE Trans. Signal Processing*, 40(8):2095–2097.

H. Peng, F. Long, and C. Ding (Aug. 2005). Feature selection based on mutual information: Criteria of max-dependency, max-relevance, and min-redundancy. *IEEE Trans. Pattern Analysis and Machine Intelligence*, 27(8):1226–1238.

L. Piegl and W. Tiller (1997). *The NURBS book*. Springer, 2nd edition.

B. Pinkowski (1993). Multiscale fourier descriptors for classifying semivowels in spectrograms. *Pattern Recognition*, 26(10):1593–1602.

M. Pitz and H. Ney (Sept. 2005). Vocal tract normalization equals linear transformation in cepstral space. *IEEE Trans. Speech and Audio Processing*, 13(5):930–944.

D. Pye and P. Woodland (Apr. 1997). Experiments in speaker normalisation and adaptation for large vocabulary speech recognition. In *Int. Conf. Acoustics, Speech, and Signal Processing*, volume 2, pages 1047–1050. Munich, Germany.

Y. Qiao, M. Suzuki, and N. Minematsu (Apr. 2009). Affine invariant features and their application to speech recognition. In *IEEE Int. Conf. Acoustics, Speech and Signal Processing*, pages 4629 –4632.

L. Rabiner and B. H. Juang (1993). *Fundamentals of Speech Recognition*. Prentice Hall Signal Processing Series. Prentice Hall, New Jersey, USA.

L. R. Rabiner and B. H. Juang (Jan. 1986). An introduction to hidden Markov models. *IEEE ASSP magazine*, 3(1).

J. Rademacher, M. Wächter, and A. Mertins (Sept. 2006). Improved warping-invariant features for automatic speech recognition. In *Proc. Int. Conf. Spoken Language Processing (Interspeech 2006 - ICSLP)*, pages 1499–1502. Pittsburgh, USA.

M. Reisert (Mar. 2008). *Group integration techniques in pattern analysis : a kernel view*. Ph.D. thesis, Albert-Ludwigs-Universität Freiburg im Breisgau, Freiburg, Germany.

H. Reitboeck and T. P. Brody (Aug. 1969). A transformation with invariance under cyclic permutation for applications in pattern recognition. *Information and Control*, 15(2):130–154.

T. Sainath, B. Ramabhadran, M. Picheny, D. Nahamoo, and D. Kanevsky (Nov. 2011). Exemplar-based sparse representation features: From TIMIT to LVCSR. *IEEE Trans. Audio, Speech, and Language Processing*, 19(8):2598 –2613.

T. N. Sainath, B. Ramabhadran, and M. Picheny (Dec. 2009). An exploration of large vocabulary tools for small vocabulary phonetic recognition. In *Proc. ASRU 2009*. Merano, Italy.

T. N. Sainath, B. Ramabhadran, D. Nahamoo, D. Kanevsky, and A. Sethy (Sept. 2010). Sparse representation features for speech recognition. In *Proc. Interspeech 2010*, pages 2254–2257. Makuhari, Japan.

G. Saon, M. Padmanabhan, R. Gopinath, and S. Chen (Jun. 2000). Maximum likelihood discriminant feature spaces. In *Proc. Int. Conf. Audio, Speech, and Signal Processing*, pages 1129–1132.

G. Saon, S. Dharanipragada, and D. Povey (May 2004). Feature space gaussianization. *Proc. ICASSP 2004*, pages I–329–I–332.

R. Schlüter, A. Zolnay, and H. Ney (Sept. 2006). Feature combination using linear discriminant analysis and its pitfalls. In *Proc. Int. Conf. Spoken Language Processing (ICSLP/Interspeech)*, pages 345–348. Pittsburgh, USA.

R. Schlüter, L. Bezrukov, H. Wagner, and H. Ney (May 2007). Gammatone features and feature combination for large vocabulary speech recognition. In *Proc. IEEE Int. Conf. Acoustics, Speech, and Signal Processing*, volume 4. Montreal, Canada.

H. Schulz-Mirbach (Aug. 1992). On the existence of complete invariant feature spaces in pattern recognition. In *Proc. Int. Conf. Pattern Recognition*, volume 2, pages 178–182. Hague, Netherlands.

H. Schulz-Mirbach (1995a). Anwendung von Invarianzprinzipien zur Merkmalgewinnung in der Mustererkennung. TR-402-95-018, Universität Hamburg, Hamburg, Germany.

H. Schulz-Mirbach (1995b). Invariant features for gray scale images. In *Mustererkennung 1995, 17. DAGM-Symposium*, pages 1–14. Springer, London, UK.

F. Seide, G. Li, and D. Yu (Sept. 2011). Conversational speech transcription using context-dependent deep neural networks. In *Proc. Interspeech*, pages 437–440. Florence, Italy.

A. D. Sena and D. Rocchesso (Nov. 2005). A study on using the mellin transform for vowel recognition. In *Proc. Sound and Music Conf.* Salerno, Italy.

B. J. Shannon and K. K. Paliwal (2003). A comparative study of filter bank spacing for speech recognition. In *Proc. Microelectronic Engineering Research Conf.*

M. A. Siegler and R. M. Stern (1995). On the effects on speech rate in large vocabulary speech recognition systems. In *Proc. Int. Conf. Acoustics, Speech, and Signal Procesing*, pages 612–615.

S. Siggelkow (2002). *Feature Histograms for Content-Based Image Retrieval.* Ph.D. thesis, Fakultät für Angewandte Wissenschaften, Albert-Ludwigs-Universität Freiburg, Breisgau, Germany.

R. Sinha and S. Umesh (May 2002). Non-uniform scaling based speaker normalization. In *Proc. IEEE Int. Conf. Acoustics, Speech and Signal Processing (ICASSP'02)*, volume 1, pages I–589 – I–592. Orlando, USA.

M. Slaney (1993). An efficient implementation of the patterson-holdsworth auditory filter bank. *Apple Tech. Rep. No.35.*

M. Slaney (Nov. 2011). *Auditory Toolbox Version 2.* Interval Research Corporation, Palo Alto, USA.

G. Stemmer (2005). *Modeling Variability in Speech Recognition.* Ph.D. thesis, Technische Fakultät der Universität Erlangen-Nürnberg, Erlangen, Germany.

G. Stemmer, C. Hacker, E. Noth, and H. Niemann (Dec. 2001). Multiple time resolutions for derivatives of Mel-frequency cepstral coefficients. *IEEE Workshop on Automatic Speech Recognition and Understanding.*, pages pp. 37–40.

S. S. Stevens (May 1957). On the psychophysical law. *Psychological Review*, 64(3):153–181.

S. S. Stevens and J. Volkmann (Jan. 1936). A scale for the measurement of the psychoacoustical magnitude pitch. *J. Acoustical Society of America*, 8(3):185–190.

A. Stolcke, B. Chen, H. Franco, V. R. R. Gadde, M. Graciarena, M.-Y. Hwang, K. Kirchhoff, A. Mandal, N. Morgan, X. Lei, T. Ng, M. Ostendorf, K. Sonmez, A. Venkataraman, D. Vergyri, W. Wang, J. Zheng, and Q. Zhu (Sept. 2006). Recent innovations in speech-to-text transcription at sri-icsi-uw. *IEEE Trans. Audio, Speech, and Language Processing*, 14(5):1729 –1744.

B. H. Story (1995). *Physiologically-based speech simulation using an enhanced wave-reflection model of the vocal tract.* Ph.D. thesis, University of Iowa.

B. H. Story (May 2005). A parametric model of the vocal tract area function for vowel and consonant simulation. *J. Acoustical Society of America*, 5(5):3231–3254.

B. H. Story and I. R. Titze (July 1998). Parameterization of vocal tract area functions by empirical orthogonal modes. *Journal of Phonetics*, 26(3):223 – 260.

B. H. Story, I. R. Titze, and E. A. Hoffman (1996). Vocal tract area functions from magnetic resonance imaging. *J. Acoustical Society of America*, 100(1):537–554.

S. Stüker, C. Fügen, S. Burger, and M. Wölfel (Sept. 2006). Cross-system adaptation and combination for continuous speech recognition: The influence of phoneme set and acoustic front-end. In *Proc. Interspeech'06 (ICSLP)*, pages 512–524. Pittsburgh, USA.

M. Temerinac, M. Reisert, and H. Burkhardt (May 2007). Invariant features for searching in protein fold databases. *Int. J. Computer Mathematics (IJCM), 'Special Issue on Bioinformatics'*, 84(5):635–651.

I. R. Titze (2006). *The Myoelastic Aerodynamic Theory of Phonation*. National Center for Voice and Speech, Iowa City, USA.

K. Torkkola (2003). Feature extraction by non-parametric mututal information maximization. *Journal of Machine Learning Research*, 3:1415–1438.

L. F. Uebel and P. C. Woodland (Sept. 1999). An investigation into vocal tract length normalisation. In *Proc. 6th European Conf. Speech Communication and Technology (EUROSPEECH'99)*, pages 2527–2530. Budapest, Hungary.

S. Umesh, L. Cohen, N. Marinovic, and D. J. Nelson (Jan. 1999a). Scale transform in speech analysis. *IEEE Trans. Speech and Audio Processing*, 7(1):40–45.

S. Umesh, L. Cohen, and D. Nelson (March 1999b). Fitting the mel-scale. *Proc. Int. Conf. Acoustic, Speech & Signal Processing*.

S. Umesh, L. Cohen, and D. Nelson (2002a). Frequency warping and the Mel scale. *IEEE Signal Processing Lett.*, 9(3):104–107.

S. Umesh, S. V. B. Kumar, M. K. Vinay, R. Sharma, and R. Sinha (May 2002b). A simple approach to non-uniform vowel normalization. In *Proc. Int . Conf. Acoustics, Speech, and Signal Processing*, volume 1, pages I–517–I–520. Orlando, USA.

S. Umesh, D. Sanand, and G. Praveen (2007). Speaker-invariant features for automatic speech recognition. In *Proc. Int. Joint Conf. Artificial Intelligence*, pages 1738–1743. Hyderabad, India.

D. Van Compernolle (Aug. 2001). Recognizing speech of goats, wolves, sheep and ... non-natives. *Speech Communication*, 35:71–79.

K. Vertanen (Nov. 2011). HTK wall street journal training recipe. web resource: http://www.keithv.com/software/htk/.

A. Viterbi (Apr. 1967). Error bounds for convolutional codes and an asymptotically optimum decoding algorithm. *IEEE Trans. Information Theory*, 13(2):260–269.

M. Wagh and S. Kanetkar (Apr. 1977). A class of translation invariant transforms. *IEEE Trans. Acoustics, Speech, and Signal Processing*, 25(2):203–205.

T. C. Walters (2011). *Auditory-Based Processing of Communication Sounds*. Ph.D. thesis, University of Cambridge.

P. P. Wang and R. C. Shiau (1973). Machine recognition of printed chinese characters via transformation algorithms. *Pattern Recognition*, 5(4):303–321.

L. Welling, H. Ney, and S. Kanthak (Sept. 2002). Speaker adaptive modeling by vocal tract normalization. *IEEE Trans. Speech and Audio Processing*, 10(6):415–426.

T. Wesker, B. Meyer, K. Wagener, J. Anemüller, A. Mertins, and B. Kollmeier (Sept. 2005). Oldenburg logatome speech corpus (OLLO) for speech recognition experiments with humans and machines. In *Proc. Interspeech 2005*. Lisbon, Portugal.

T. Wesker, B. Meyer, K. Wagener, J. Anemüller, A. Mertins, and B. Kollmeier (Oct. 2008). Oldenbug logatom corpus (ollo). web resource: http://sirius.physik.uni-oldenburg.de/html/download_ollo.html.

J. G. Wilpon and C. S. Jacobsen (May 1996). A study of speech recognition for children and the elderly. In *Proc. Int. Conf. Acoustics, Speech, and Signal Processing*, pages 349–352.

S. Young, N. Russell, and J. Thornton (1989). Token passing: a simple conceptual model for connected speech recognition systems. Technical Report CUED/F-INFENG/TR38, University of Cambridge: Department of Engineering, Cambridge, UK.

S. Young, G. Evermann, M. Gales, T. Hain, D. Kershaw, X. A. Liu, G. Moore, J. Odell, D. Ollason, D. Povey, V. Valtchev, and P. Woodland (2009). *The HTK Book (for HTK Version 3.4.1)*. Cambridge University Engineering Department, Cambridge, UK.

S. J. Young, J. J. Odell, and P. C. Woodland (1994). Tree-based state tying for high accuracy acoustic modelling. In *Proc. Workshop Human Language Technology*, HLT '94, pages 307–312. Association for Computational Linguistics, Stroudsburg, USA.

Y. Zhang and J. Zhou (May 2004). Audio segmentation based on multi-scale audio classification. *Proc. Int. Conf. Acoustics, Speech, and Signal Processing*, 4:iv–349–iv–352.

Mathematical Notation

The following mathematical notation has been used throughout this work:

General	
$\boldsymbol{X}$	matrix $\boldsymbol{X}$
$\boldsymbol{X}^{\mathrm{T}}$	transpose of $\boldsymbol{X}$
$\boldsymbol{X}^{-1}$	inverse of $\boldsymbol{X}$
$\lvert\boldsymbol{X}\rvert$	determinant of $\boldsymbol{X}$

Features and Feature Sequences	
$\boldsymbol{y}_t$	feature vector $\boldsymbol{y}_t \in \mathbb{R}^d$ at time t
$\boldsymbol{Y}_{1:T}$	sequence of feature vectors, $\boldsymbol{Y}_{1:T} \in \mathbb{R}^{d \times T}$, where $\boldsymbol{Y}_{1:T} = \begin{bmatrix}\boldsymbol{y}_1 & \boldsymbol{y}_2 & \dots & \boldsymbol{y}_T\end{bmatrix}$
$T^{(r)}$	number of feature vectors of vector sequence r, $T^{(r)} \in \mathbb{N}^+$
$\boldsymbol{y}_t^{(r)}$	feature vector $\boldsymbol{y}_t \in \mathbb{R}^d$ at time t of utterance r
$\boldsymbol{Y}^{(r)}$	r-th feature vector sequence $\boldsymbol{Y}^{(r)} = \begin{bmatrix}\boldsymbol{y}_1^{(r)} & \boldsymbol{y}_2^{(r)} & \dots & \boldsymbol{y}_{T^{(r)}}^{(r)}\end{bmatrix}$
R	number of all available utterances
$\boldsymbol{x}_t$	"clean" speech feature vector at time t
$\boldsymbol{n}_t$	"noise" feature vector at time t

HMM Parameters

Q	number of states of a HMM
S_j	state j of a HMM
$\mathcal{A}$	set of all transition probabilities
a_{ij}	transition probability for changing from state i to state j
$b_j(\boldsymbol{y})$	observation probability distribution assigned to state j
$c^{(jm)}$	a priori probability of Gaussian component m of state j
$\boldsymbol{\mu}^{(jm)}$	mean of Gaussian component m of state j
$\mu_i^{(jm)}$	element i of mean vector $\boldsymbol{\mu}^{(jm)}$ of state j
$\boldsymbol{\Sigma}^{(jm)}$	covariance matrix of Gaussian component m of state j
$\mathcal{N}(\boldsymbol{\mu}, \boldsymbol{\Sigma})$	Gaussian distribution with mean $\boldsymbol{\mu}$ and covariance matrix $\boldsymbol{\Sigma}$
$\mathcal{N}(\boldsymbol{y}_t; \boldsymbol{\mu}, \boldsymbol{\Sigma})$	likelihood of observation $\boldsymbol{y}_t$ being generated by $\mathcal{N}(\boldsymbol{\mu}, \boldsymbol{\Sigma})$
$\mathcal{B}$	set of all feature vector probability distributions
$\boldsymbol{\lambda}$	set of HMM parameters $\boldsymbol{\lambda} = \{\boldsymbol{A}, \mathcal{B}\}$
$p(\boldsymbol{Y}_{1:T} \mid \boldsymbol{\lambda})$	likelihood of observation sequence $\boldsymbol{Y}_{1:T}$ given model parameters $\boldsymbol{\lambda}$
θ_t	state occupied at time t

Language Models

q	a phone such as /a/, /t/, etc
w	word w
$\boldsymbol{W}_{1:P}$	word sequence $\boldsymbol{W}_{1:P} = (w_1, w_2, \ldots, w_P)$ of length P
$C(\boldsymbol{W}_{1:P})$	number of occurrences of word sequence $\boldsymbol{W}_{1:P}$
$P(\boldsymbol{W}_{1:P})$	probability of word sequence $\boldsymbol{W}_{1:P}$
ω_k^W	word class of word w_k

Feature Transforms

L	number of classes
ω_i	class i
M	number of Gaussian densities
N	number of all feature vectors
N_i	number of all feature vectors belonging to class i
$\mathcal{C}_i$	set of all indices whose feature vectors belong to class i
$\boldsymbol{\mu}_G$	(global) mean of all feature vectors
$\boldsymbol{\Sigma}_G$	(global) covariance matrix of all feature vectors
c_i	a priori probability of class i
$\boldsymbol{\mu}_i$	mean of all feature vectors belonging to class i
$\boldsymbol{\Sigma}_i$	covariance matrix of all feature vectors belonging to class i
p	target dimensionality, $p \in \mathbb{N}^+$
$\boldsymbol{A}$	linear transform, $\boldsymbol{A} \in \mathbb{R}^{p \times d}$, $p < d$

Feature Extraction

$x(m)$	discrete-time signal x
τ_l	frame length in seconds, $\tau_l \in \mathbb{R}^+$
τ_s	frame shift in seconds, $\tau_s \in \mathbb{R}^+$
τ_l^d	frame length in number of samples, $\tau_l^d \in \mathbb{N}^+$
τ_s^d	frame shift in number of samples, $\tau_s^d \in \mathbb{N}^+$
K	number of subbands of time-frequency representation
$\boldsymbol{f}_c$	filter center frequencies $\boldsymbol{f}_c = (f_c(1), f_c(2), \ldots, f_c(K))$
s_r	sampling rate in Hz
$\boldsymbol{s}_n$	vector of spectral values at time index n, $\boldsymbol{s}_n \in \mathbb{R}^K$
$s_n(k)$	spectral value at time index n and channel index k
$\xi(f)$	auditory scale value for frequency f
$\boldsymbol{T}(\boldsymbol{x})$	invariant transform $\boldsymbol{T}$
$\mathcal{I}_{\boldsymbol{T}}(\boldsymbol{x})$	set of invariants of $\boldsymbol{x}$ for a given invariant transform $\boldsymbol{T}$
$\mathcal{O}(\boldsymbol{x})$	orbit of x, that is the set of all equivalent observations for $\boldsymbol{x}$
$S(n, \omega)$	short-time discrete-time Fourier transform

B

Corpora

Speech corpora provide standardized sets of speech utterances together with different types of ground truth informations for these utterances. Corpora provide the possibility for comparing the performance of different methods on the same database between research and development groups. The experiments of this work were conducted on corpora that are widely used within the ASR research community and allow for focusing on different kinds of variabilities. As is explained in more detail in the following, these variabilities are the vocal tract length, the noise condition, and the rate of speech.

B.1. TIMIT

The *Texas Instruments/Massachusetts Institute of Technology* (TIMIT) corpus consists of read speech utterances to provide speech data for the acquisition of acoustic-phonetic knowledge and for the evaluation of ASR systems (Garofolo et al., 1993). TIMIT contains 6300 utterances in total with a sampling rate of 16 kHz. Overall, 630 speakers from eight major dialect divisions of the United States spoke ten sentences. There are 438 (70%) male speakers and 192 (30%) female speakers. The corpus contains about five hours of speech.

Table B.1.: Overview of the speech material of the TIMIT corpus.

sentence type	# sentences	# speakers/ sentence	total	# sentences/ speakers
dialect (SA)	2	630	126	2
compact (SX)	450	7	3150	5
diverse (SI)	1890	1	1890	3
total	2342	-	6300	10

Corpus text material The sentence types of TIMIT are divided into three types: dialect sentences, phonetically compact sentences, and phonetically-diverse sentences. Within the corpus documentation these types are abbreviated as SA, SX, and SI sentences, respectively. The SA sentences are supposed to expose the dialectal variations of the speakers and were read by all 630 speakers. The two dialect sentences are "She had your dark suit in greasy wash water all year" and "Don't ask me to carry an oily rag like that". The SX sentences should provida a good coverage of pairs of phones. The phonetically diverse sentences were selected from existing test sources and a collection of dialogs from stage plays. Table B.1 gives an overview of the number of sentences of each type, the number of speakers per sentence of a given type, and the number of sentences per speaker for each type.

Data sets The training set consists of the 462 speakers that are not included in either the core or the complete test sets. Excluding the SA sentences the training set contains 3696 utterances. There is no overlap between the texts of the training and the test utterances. A detailed list of the speakers for each of the described sets can be found in the work of Halberstadt and Glass (1998). Within this work, additional segmentations of the training and test data were used to allow for an evaluation of the feature types under mismatching training-test conditions with respect to different average VTLs. This is achieved by using only male utterances from the training set, and only female data from the test set. This scenario is denoted as "M-F" within the according tables of this work. Similarly, "F-M" denotes the training on female utterances from the training set and test on the male utterances from the test set. If the standard training and test sets are used, this matching scenario is also denoted as "FM-FM" in the tables.

The TIMIT core test set consists of two male and one female speaker from each of the eight dialect regions. There are five SX and three SI sentences spoken by each speaker, for a total of 192 utterances in the core test set.

The complete test set included all seven repetitions of the SX sentences. This results in the addition of another 144 speakers to the core test set. Overall, there are 168 speakers, each uttering eight sentences. Similar to the core test set, the SI sentences were also discarded in the complete test set. This results in 1344 utterances in total for the complete test set. The described data sets are sometimes also referred to as "NIST complete test set" and "NIST training set" (Halberstadt and Glass, 1998).

Phones and phone classes The TIMIT corpus provide a time-aligned word transcription and a time-aligned phonetic transcription for each utterance. Overall, the phonetic set consists of 61 phones, which are shown as ARPAbet symbols in Table B.2. Following a standard procedure (Lee and Hon, 1989), the 61 TIMIT labels were collapsed into 39 labels before scoring and the glottal stops (/q/), were ignored. The mapping is shown in Table B.3.

Comparable Results in this Work The accuracies given in the tables listed in the following were obtained from the same ASR system and, thus, result from the same acoustic and language modeling. In these tables, accuracies are reported for different feature types and different target dimensionalities, as well as accuracies when VTLN and/or MLLR was used or not used:

- Table 3.1 on page 64
- Table 4.9 on page 108
- Table 4.10 on page 110
- Table 4.11 on page 111
- Table 4.12 on page 112
- Table 4.14 on page 116
- Table 4.17 on page 119

B.2. TIDigits and AURORA 2

Texas Instruments collected a corpus of speech for the purpose of "designing and evaluating algorithms for speaker-independent recognition of connected digit sequences" (Leonard, 1984) and is called "*TIDigits*". The utterances are provided with a sampling rate of 20 kHz. For the experiments within this work, the utterances were downsampled to 16 kHz. Overall, the corpus contains read utterances

Table B.2.: The 61 ARPAbet symbols for the phones contained in the TIMIT transcriptions together with exemplary occurences (Garofolo et al., 1993; Rabiner and Juang, 1993).

ARPAbet	Example	ARPAbet	Example
aa	b*o*tt	ix	deb*i*t
ae	b*a*t	iy	b*ee*t
ah	b*u*t	jh	*j*oke
ao	b*ough*t	k	*k*ey
aw	b*ou*t	kcl	k closure
ax	*a*bout	l	*l*ay
ax-h	s*u*spect	m	*mom*
axr	butt*er*	n	*noon*
ay	b*i*te	ng	si*ng*
b	*b*ee	nx	wi*nn*er
bcl	b closure	ow	b*oa*t
ch	*ch*oke	oy	b*oy*
d	*d*ay	p	*p*ea
dcl	d closure	pau	pause
dh	*th*en	pcl	p closure
dx	mu*dd*y	q	bat (glottal stop)
eh	b*e*t	r	*r*ay
el	bott*le*	s	*s*ea
em	bott*om*	sh	*sh*e
en	butt*on*	t	*t*ea
eng	wash*ing*ton	tcl	t closure
epi	epenthetic silence	th	*th*in
er	b*ir*d	uh	b*oo*k
ey	b*ai*t	uw	b*oo*t
f	*f*in	ux	t*oo*t
g	*g*ay	v	*v*an
gcl	g closure	w	*w*ay
hh	*h*ay	y	*y*acht
hv	a*h*ead	z	*z*one
ih	b*i*t	zh	a*z*ure
h#	initial and final silence of utterance		

Table B.3.: Mapping from the 61 original TIMIT labels to 39 classes according to Lee and Hon (1989).

class	phones	class	phones
1	iy	20	n en nx
2	ih ix	21	ng eng
3	eh	22	v
4	ae	23	f
5	ax ah ax-h	24	dh
6	uw ux	25	th
7	uh	26	z
8	ao aa	27	s
9	ey	28	zh sh
10	ay	29	jh
11	oy	30	ch
12	aw	31	b
13	ow	32	p
14	er axr	33	d
15	l el	34	dx
16	r	35	t
17	w	36	g
18	y	37	k
19	m em	38	hh hv
39	bcl pcl dcl tcl gcl kcl q epi pau h#		

from 326 speakers that came from 21 different dialectical regions of the United States. The number of speakers and the age range for each of the the categories man, woman, boy, and girl are shown in Table B.4.

Corpus text material The utterances collected from the speakers are continuous English digit sequences and were spoken into a close-talking microphone. The set of digits was "zero", "one", "two", . . ., "nine", "oh", and the distribution of each digit over all sequences was made uniform. Each of the speakers uttered 77 digit sequences. Six different types of sequences were recorded, which are shown in Table B.5. Because of the very limited vocabulary of the TIDigits task, whole-word models were used for acoustic modeling during the experiments. The number of states was chosen proportional to the average length of the individual digit utterances and are listed in Table B.6.

Table B.4.: Number of speakers and according age ranges for the speaker groups of TIDigits

category	number	age range [years]
man	111	21-70
woman	114	17-59
boy	50	6-14
girl	51	8-15

Table B.5.: Sequence types contained in TIDigits

number	sequence type
22	isolated digits (two tokens of each of the eleven digits)
11	two-digit sequences
11	three-digit sequences
11	four-digit sequences
11	five-digit sequences
11	seven-digit sequences

Data sets TIDigits provides predefined segmentations into training and test set. The segmentation was based on speaker category and dialect classification. This procedure yielded two sets, each containing approximately half the speakers of each category and dialect classification. Overall, the training and the test set consist of 12549 and 12547 utterances, respectively.

AURORA 2 The AURORA 2 corpus (Hirsch and Pearce, 2000) is based on the utterances of TIDigits and represents a task that focuses on on the recognition of continuous digit sequences under noisy conditions and limited bandwidth. Therefore, the TIDigits utterances were downsampled to 8 kHz. To simulate the transmission characteristics of typical telecommunication equipment, the utterances were convolved with standardized ITU filters (ITU, 2001) to simulate the transmission characteristics of typical communication terminals. Noise signals that consist of stationary and also of nonstationary noises were added with SNRs between -5 to +20 dB. Two training modes are defined for the AURORA task:

- training on clean speech only ("clean speech training"), and

Table B.6.: Number of states for each model used for TIDigits experiments.

word	number of states
silence	11
one	14
two	12
three	14
four	14
five	16
six	17
seven	16
eight	13
nine	15
zero	17
oh	13

- training on clean and noisy speech ("multi-style training"), which is the preferred method in practice.

In both modes, the training set consists of 8440 utterances from the TIDigits training set. In case of multi-style training, 20 equally sized subsets were generated, each representing different noise and SNR conditions. For testing, three test sets with different characteristics are defined. Test set A provides a high match (with respect to the used noise types) between training and test data. Test set B uses noises that do not match with the noise conditions from the training data. Finally, test set C simulates a mismatch between the transmission characteristics of the telecommunication terminals used in the training and the test sets (Hirsch and Pearce, 2000). For the experiments within this work, the average accuracies of the three test sets are reported for both, clean-speech and multi-style training.

Comparable Results in this Work The accuracies given in the tables listed in the following were obtained from the same ASR system and, thus, result from the same acoustic and language modeling. In these tables, accuracies are reported for different feature types and noise conditions:

- Table 3.2 on page 65
- Table 5.1 on page 124
- Table 5.2 on page 125

- Table 5.3 on page 129
- Table 5.4 on page 136
- Table 5.5 on page 137

B.3. OLLO

The *Oldenburg Logatome* (OLLO) speech corpus (Wesker et al., 2005) has been designed for better modeling, feature extraction and adaptation in the presence of intrinsic variabilities. The current version of the corpus, OLLO 2.0 (Wesker et al., 2008), is a multi-lingual database that contains read speech of 50 different speakers. The speaker-dependent variabilities contained in the utterances are gender (25 male, 25 female), age (ranging from 10 to 64), and region (Bavarian, East Frisian, Eastphalian, Belgium). Six different speaker-independent pronunciation variabilities have been recorded for each logatome: speaking rate (fast, normal, slow), speaking effort (normal, questioning), and speaking style (normal, soft, low). Besides the logatomes, different sentences and monosyllabic words have been recorded in normal speaking style. However, only the logatome utterances were used in the experiments of this work.

Corpus text material Either consonant-vowel-consonant or vowel-consonant-vowel combinations ("logatomes") were read by the speakers. The inital and final phoneme were the same in each logatome. Overall, OLLO provides 150 logatomes. Each logatome was read by each speaker three times and each of the six pronunciation variabilities was considered. Hence, the complete corpus provides 135.000 logatome utterances with a total length of about 54 hours.

Data sets The OLLO corpus provides two predefined segmentations for the definition of training and test sets. According to the documentation of the corpus, variant "A" is designed for speaker-independent systems for ASR, while variant "B" is optimized for speaker-dependent ASR experiments. Within this work, variant "A" was used. In this variant, five speakers of each dialect region are contained in the training and test set. In the experiments described in this work, only the logatomes that are assigned to one of the groups "normal", "slow", or "fast" were considered in order to focus on the rate-of-speech variability. Hence, the used training set contains 26474 utterances and the test set contains 29510 utterances.

Comparable Results in this Work The accuracies given in the tables listed in the following were obtained from the same ASR system and, thus, result from the same acoustic and language modeling. In these tables, accuracies are reported for different feature types and speaking rates:

- Table 3.3 on page 66
- Table 4.16 on page 117

C
Additional Results

C.1. Parameters and Results for Section 4.3

In Section 4.3 experiments are conducted which evaluate different scales for locating the frequency centers of a filter bank. Table C.1 shows the parameters for each considered scale of these experiments. Besides the different scales, the experiments evaluated triangular-shaped and gammatone filters, as well as a scaling of the bandwidth of the filters. All results for the experiments about the optimization of warping schemes as described in Section 4.3 are illustrated in Figure C.1.

Table C.1.: List of parameters of all considered scales during the experiments described in Section 4.3.

Scale name	x_1	x_2	w
	6985	41	2.0
	6907	39	1.9
	6829	38	1.9
	6750	37	1.8
	6672	36	1.8
	6594	35	1.7
	6516	34	1.7
log	6438	34	1.6
	6369	109	1.6
	6301	185	1.6
	6232	261	1.5
	6164	337	1.5
	6096	413	1.4
	6027	489	1.4
	5959	565	1.3
ERB	5891	643	1.3
	5839	718	1.3
	5787	795	1.3
	5736	873	1.2
	5685	950	1.2
	5633	1027	1.2
	5582	1104	1.2
	5530	1181	1.2
mel	5479	1258	1.2
	5428	1335	1.1
	5376	1412	1.1
	5325	1489	1.1
	5273	1566	1.1
	5222	1643	1.1
	5170	1720	1.1
	5119	1797	1.0
	5068	1874	1.0

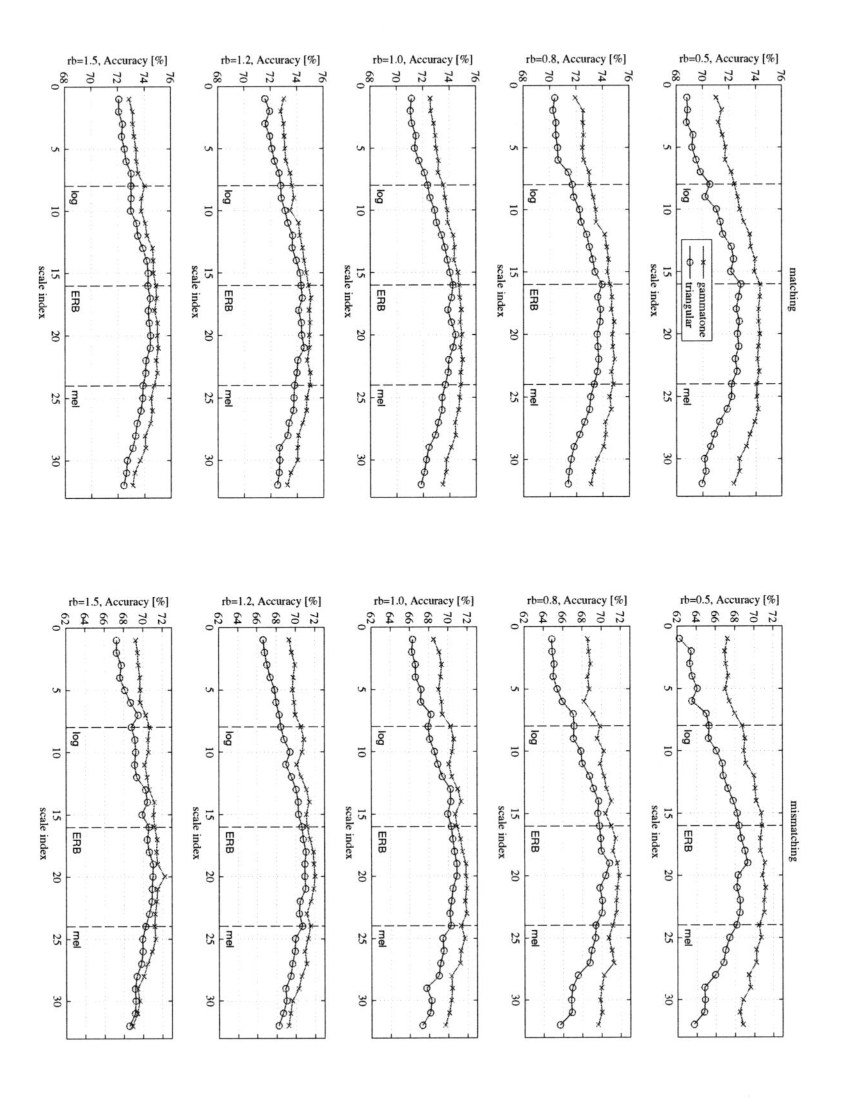

Figure C.1.: Results for the experiments about the optimization of warping schemes as described in Section 4.3. "rb" denotes the scaling factor for the bandwidth of the filters in the individual filter banks.

D

IIFCopy: A C++ Implementation for the Computation of Invariant Integration Features

For the computation of invariant integration features (IIF) as described in Section 4.6 a C++ version was implemented as part of this work. With the training and test sets usually comprising thousands of utterances one motivation for this implementation was the decrease in computational time in comparison to the corresponding Matlab implementation. Furthermore, having a possibility to compute integration features (possibly in parallel) without the need of Matlab was another motivation.

Besides the computation of the features themselves, the implemented console application also allows for the computation of feature transforms with PCA or LDA, as well as for the parameter selection for IIFs with a given training set. Binaries for various Linux as well as Windows platforms were compiled. A screenshot of the help screen of the application is shown in Figure D.1.

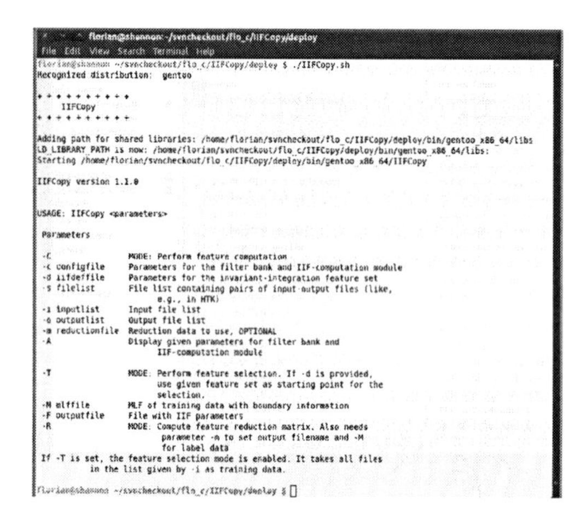

Figure D.1.: Screenshot of the console application “IIFCopy”.

German Translations of Title, Introduction, and Summary

E.1. Titel

Invarianz-Merkmals-Extraktions-Methoden und Verbesserte Vokaltraktlängen-Normalisierungsmethoden zur Automatischen Spracherkennung

E.2. Einführung

In unserem täglichen Leben führen wir auf leichte Weise Konversationen mit vielen verschiedenen Leuten und unter vielen verschiedenen Bedingungen. Die Fähigkeit zum Hören und Verstehen von menschlicher Sprache ist ein zentrales Element dieser Interaktivität. Diese hat sich über Millionen von Jahren entwickelt und die notwendigen Aktionen für diese Interaktivität werden typischerweise unbewusst ausgeführt. Diese Arbeit handelt von automatischer Spracherkennung (engl. automatic speech recognition, ASR). Als Problemstellung hat ASR das Übersetzen eines Sprachsignals in eine entsprechende textliche Darstellung durch eine Maschine. Es ist Teil des weiten Feldes der Sprachverarbeitung und eine zentrale Komponente in vielen Anwendungen, die mittlerweile im Alltag aufgekommen sind. Beispiele sind die automatische Übersetzung von Nachrichten, Informationsgewinnung, oder Sprachtranslation (Feng et al., 2012).

Lippmann (1997) verglich die Leistung menschlichen und maschinellen Verstehens von Sprache auf sechs verschiedenen Sprachdatensätzen, sogenannte Korpora, mit Wortschätzen von 10 bis hin zu mehr als 65000 Worten und Sprecharten von "isolierten Worten", "vorgelesener Sprache" und "spontaner Sprache". Er zeigte, dass die menschliche Leistungsfähigkeit zur Spracherkennung bei weitem die entsprechende maschinelle Leistungsfähigkeit übertrifft und meist um Größenordnungen besser ist, als die der damaligen Erkennungssysteme in ungestörter wie auch in gestörter Umgebung. Benzeghiba et al. (2007) argumentiert, dass, obwohl die Experimente von Lippmann vor mehr als einem Jahrzehnt durchgeführt wurden, der Abstand zwischen der maschinellen und der menschlichen Erkennungsleistung immer noch substantiell ist. Im Hinblick auf gegenwärtige ASR-Systeme, die immer häufiger im Alltag auftreten, weist Feng et al. (2012) darauf hin, dass viele interessante Herausforderungen in diesem Gebiet immer noch zu lösen sind oder gar erst entstehen. So sind sprecher-unabhängige Merkmalsextraktion, Rauschrobustheit und auditorische Repräsentationen aktuelle Forschungsfelder.

Im Allgemeinen sind menschliche Hörer in der Lage, Sprache zuverlässig in einer Vielzahl unterschiedlichen von Umgebungen und unterschiedlichen sprecherspezifischen Variabilitäten zuverlässig zu verstehen. Die Größe der Sprecher stellt eine dieser Variabilitäten dar. Das Sprachsignal erreicht normalerweise das auditorische System eines Hörers in Form einer akustischen Welle und durchwandert mehrere Verarbeitungsebenen, so dass die Nachricht, die in dem Signal enhalten ist, verstanden werden kann. Dies bedeutet, dass die Repräsentation des Sprachsignals innerhalb des akustischen Systems zu einem bestimmten Zeitpunkt derart transformiert wird, dass das Ergebnis dieser Transformation unabhängig von den Variabilitäten, die einen störenden Einfluss auf das Sprachsignal haben, ist. Das "Front-end" eines Spracherkennungs-Systems transformiert ebenfalls das Eingangs-Sprachsignal. Es parametrisiert das Signal in eine Darstellung, die für eine nachfolgende Erkennung durch eine Maschine geeignet ist. Diese Ebene wird auch als Merkmalsextraktions-Ebene (engl. feature extraction stage) bezeichnet. Momentane Standard-Merkmalsextraktionsmethoden für ASR behandeln die Effekte unterschiedlich großer Sprecher typischerweise nicht direkt auf der Merkmalsextraktions-Ebene, sondern auf der Ebene der akustischen Modelle und des Dekodierens. Die Methoden, die in dieserArbeit beschrieben werden, konzentrieren sich auf die Merkmalsextraktions-Ebene eines Spracherkennungs-Systems. Die involvierten Methoden machen dabei von sogenannten "Invarianztransformationen" Gebrauch, um Robustheit gegenüber Variabilitäten zu erreichen, die durch unterschiedliche Sprecher hervorgerufen werden. Auch werden Methoden vorgestellt und untersucht, die die Robustheit gegenüber gestörten Umgebungsbedingungen erhöhen. In einem weiteren Kapitel werden Ansätze zum Schätzen der Spektraleffekte durch unterschiedliche Vokaltraktlängen vorgestellt und zur Verbesserung einer nachfolgenden Erkennung verwendet.

Die Forschung im Gebiet der Spracherkennung ist interdisziplinär und umfasst mehrere Felder, die unter anderem die Signalverarbeitung, Optimierung, maschinelles Lernen und statistische Methoden, sowie Linguistik, Psychoakustik und Phonologie beinhaltet. Die nächsten beiden Abschnitte geben einen Überblick über die menschliche Sprachproduktion und -wahrnehmung, sowie einen Überblick über die Variabilitäten, die im Kontext der Spracherkennung beobachtet werden können und, je nach Anwendungsfall, berücksichtigt werden sollten. Am Ende dieses Kapitels wird ein Überblick über die wissenschaftlichen Beiträge dieser Arbeit gegeben.

E.2.1. Prozesse der Sprachproduktion und -wahrnehmung

Gesprochene Sprache wird verwendet, um Informationen von einem Sprecher hin zu einem Zuhörer zu kommunizieren (Huang et al., 2001b). Der Prozess der Generierung und Verarbeitung von Sprache kann als Sequenz von Aktivitäten beschrieben werden wie es in Darstellung 1.1 gezeigt ist. Obwohl nicht explizit gezeigt, verwendet der Sprecher seine eigene Sprache auch als Rückmeldung, um den Spracherzeugungs-Prozess zu kontrollieren.

Schematischer Überblick über die Sprachproduktion und -wahrnehmung

Der Produktionsprozess wird initiiert durch die Intention des Sprechers Informationen durch Sprache zu übermitteln. Eine Nachricht muss dafür zunächst mental formuliert werden und umfasst semantisches und pragmatisches Wissen. Nach der Formulierung muss die Nachricht in eine Sequenz von Phonemen konvertiert werden. Ein Phonem kann als ideale Lauteinheit mit einer kompletten Menge von artikulatorischen Gesten beschrieben werden (Deller et al., 1993). Bestimmte Phonemgruppen stellen Wörter dar, womit wiederum ganze Sätze durch Aneinanderreihung gebildet werden. Der Sprecher muss entscheiden, welche Sprache er verwenden will und muss daher lexikalisches und syntaktisches Wissen über die entsprechende Sprache besitzen. Auch muss die Sprecher eine bestimmte Prosodie wählen. Im Allgemeinen muss der Sprecher einen Sprachkode für die intendierte Nachricht wählen. Dann hat der Sprecher eine Sequenz von neuromuskularen Aktionen auszuführen, um Stimmbandvibrationen, Luftdruck, Vokaltraktformen, Lippen-, Kiefer-, Zungen- und Velumbewegungen mit adequater Zeiteinteilung zu produzieren. Die resultierende akustische Welle, die der Sprecher emittiert, wird über einen Kanal übertragen.

Während der Übertragung treten verschiedene Arten von Störungen auf, die auf das Sprachsignal einwirken. Wenn das Sprachsignal den Hörer erreicht, wird es zur Basilarmembran im Innenohr übermittelt. Eine nichtlineare Spektralanalyse wird durchgeführt und deren Ausgabe wird in Aktivitätsignale im auditorischen Nerv umgewandelt. Diese Signale werden in einen Sprachkode auf höheren Ebenen des Gehirns des Zuhörers umgewandelt. Durch die Verwendung semantischen und pragmatischen Wissens wird ein Verständnis für die Nachricht bei dem Zuhörer erreicht.

Artikulatoren und Modell Wie oben erwähnt, ist die Sprachproduktion ein dynamischer Prozess, der mehrere anatomische Strukturen involviert. Ein Überblick über diese Strukturen ist in Darstellung 1.2 gegeben. Die Lungen erzeugen einen Luftdruck, der zu einem Luftfluss durch die Luftröhre, sowie die Rachen-, Mund- und Nasenhöhle führt. Die Rachen- und die Mundhöhle gemeinsam werden für gewöhnlich als Vokaltrakt bezeichnet. Der Kehlkopf beinhaltet die Stimmbänder und die Glottis. Die Länge des Vokaltraktes, gemessen als die Distanz zwischen der Glottis und der Lippen, wird im Folgenden als Vokaltraktlänge (engl. vocal tract length, VTL) bezeichnet. Der Vokaltrakt beinhaltet das Velum, die Zunge und die Zähne und endet bei den Lippen. Velum, Zunge, Zähne und Lippen zusammen werden auch als Artikulatoren bezeichnet. Aus Ingenieurspersepktive stellt der Vokaltrakt ein akustisches Filter dar, das einen bedeutenden Anteil bei der Produktion von Lauten trägt. Die Eigenschaften des Systems werden durch die Artikulatoren kontrolliert. Um ein stimmhaften Sprachsignal zu erzeugen, wird Luft durch die Glottis gezwungen, so dass die Stimmbänder anfangen zu vibrieren und dabei periodische Pulse produzieren. Dies führt zu harmonischen Strukturen im Spektrum des stimmhaften Lautes. Die Rate f_0, mit der sich die Stimmbänder öffnen, wird Fundamentalfrequenz oder auch (engl.) Pitch genannt. Wie von Deller et al. (1993, p. 114) hingewiesen, wird der Begriff Pitch im Bereich der Psychoakustik oft auch als wahrgenommene Fundamentalfrequenz verwendet. Stimmlose Laute werden ohne eine Schwingung der Stimmbänder generiert und haben daher keine harmonischen Strukturen in deren Spektren, sondern können vielmehr durch bandbegrenztes Rauschen charakterisiert werden.

Ein grundlegendes zeit-diskretes Modell der Sprachproduktion ist in Darstellung 1.3 gezeigt. Neben einem Vokaltraktmodell $H(z)$, wird hier ein Abstrahlungsmodel $R(z)$ gezeigt, welches die akustischen Eigenschaften des Übergangs vom Vokaltrakt zur Umgebung modelliert. Das gesamte System wird durch ein Anregungssignal $u(m)$ angeregt. Im Falle von stimmlosen Sprachintervallen besteht das Anregungssignal aus Rauschen. Eine Sequenz von glottalen Impulsen wird hingegen bei stimmhaften Lauten verwendet. Die Pulse werden durch ein Glottis-Modell $G(z)$ beschrieben, welches durch eine Pulsfolge mit bestimmtem Pitch P_0 getrieben wird. Die Amplitude des produzierten Signals wird durch einen Skalierungsfaktor beeinflusst.

E.2.2. Variabilitäten Menschlicher Sprache

Im Allgemeinen erreichen heutige ASR-Systeme nicht die Erkennungsleistung des Menschen. Es ist möglich, ein Erkennungssystem für einen bestimmten Anwendungsfall in bestimmtem Kontext und für einen bestimmten Sprecher zu konstruieren, welches eine hohe Erkennungsleistung aufweist. Durch den hohen Grad an Variabilität jedoch, den Sprache von Natur aus enthält, ist es heutzutage immer noch eine Herausforderung, ein Erkennungssystem zu konstruieren, welches in der Lage ist, eine beliebige Sprache unter beliebigen Bedingungen von jedem Sprecher zu erkennen. Einen umfassenden Überblick über die Variabilitäten, mit denen ein ASR-System umgehen muss, wird in der Arbeit von Benzeghiba et al. (2007), sowie in der Thesis von Stemmer (2005) gegeben. Eine übergreifende Klassifizierung der Hauptvariabilitäten wird auch von Huang et al. (2001b, pp. 414-417) gegeben. Eine Zusammenfassung der verschiedenen Typen von Variabilitäten ist im Folgenden gegeben. Da keine einheitliche Gruppierung der Variabilitäten im Feld der automatischen Spracherkennung existiert, wurde im Folgenden eine eigene Gruppierung in vier Kategorien gewählt.

Umgebungs-Variabilitäten Die meisten Spracherkennungs-Anwendungen beziehen eine Umgebung mit ein, die verschiedene Arten von Lautquellen beinhaltet. Beispielsweise könnten redende Menschen im Hintergrund sein, Maschinen Lärm emittieren oder Klingeln läuten. Innerhalb eines Raumes könnte ein großer Nachhall existieren. Die störende Lärmquelle kann an einer Position fixiert sein oder sich bewegen und dadurch die spektralen Charakteristiken des Störgeräuschs verändern. Die Charakteristik des Rauschsignal oder des zu erkennenden Sprachsignals könnten sich im Laufe der Zeit ändern. Im Falle von Telefonsprache könnten Störgeräusche auch von unterschiedlichen Quellen ausgehen, zum Beispiel vom Übertragungskanal oder der Analog-Digital-Konvertierung. Die Mikrofone eines ASR-Systems werden all diese Geräusche zusätzlich zur eigenlichen Sprache aufnehmen.

Variabilitäten der Sprechweise Abhängig von der Anwendung oder der Situation artikulieren menschliche Sprecher auf unterschiedliche Weisen, welche jeweils unterschiedliche Charakteristika aufweisen. Erkennungssysteme müssen mit all diesen Charakteristika umgehen können, wenn sie unter den entsprechenden Bedingungen eingesetzt werden sollen. Falls ein ASR-System ausschließlich isolierte Worte erkennen soll, zum Beispiel im Falle eines Kommandowort-Erkenners, so stellen die Pausen zwischen den individuellen Kommandos deutliche Grenzen mit Stille-Segmenten dar. Falls jedoch gelesene Sprache verarbeitet werden soll, wie im Falle von Nachrichtentexten oder Diktieranwendungen, so wird ein kontinuierliches Spracherkennungssystem benötigt. Der Vorteil eines ASR-Systems, welches in der Lage ist, kontinuierliche Sprache zu verarbeiten, ist die Möglichkeit einer natürlicheren Interaktivität mit den menschlichen Benutzern. Alleva et al. (1998) berichten, dass Benutzer dazu tendieren, zu isolierter Sprache zu wechseln, wenn sie den Einruck haben, dass das ASR-System den Benutzer nicht versteht. Allerdings wird isolierte Sprache von kontinuierlichen Spracherkennern nicht so gut erkannt, wie es ein dedizierter Erkenner für isolierte Worte tun würde. Einige Gründe dafür sind die unterschiedlichen Arten der Artikulation und der Sprechraten. Gelesene Sprache und dieser Art ähnliche Sprachtypen sind für gewöhnlich deutlich und wohl artikuliert.

Eine der Herausforderungen im Bereich der ASR-Forschung heutzutage ist die Erkennung von informeller, spontaner Sprache (Benesty et al., 2008) wie zum Beispiel Mensch-zu-Mensch-Konversationen. Diese Art von Sprache ist oft nicht klar artikuliert, Schmatzen durch die Lippen kann auftreten und Füllwörter wie (engl.) "uh" oder "well", so wie umgangssprachliche Ausdrücke wie (engl.) "she's out" anstelle von "she is going out" oder "hey guys" anstelle von "hello, sir" können auftreten. Oftmals können Verschleifungen von bestimmten Phonemen, Silben oder sogar größeren Sprachsegmenten beobachtet werden. Die Variabilität zeitlicher und spektraler Strukturen spontaner Sprache ist größer im Vergleich zu vorgelesener Sprache (Benzeghiba et al., 2007). Die Sprechrate (engl. *rate of speech*, ROS)[1] ist eine Hauptvariabilität mit der ein Erkennungssystem für spontane Sprache umgehen muss. Die ROS ist typischerweise eine Variabilität, die sich mit den verschiedenen Sprecharten ändert und einen entsprechenden Einfluss auf die Erkennungsrate eines ASR-Systems hat: Je größer die ROS, desto höher ist die Fehlerrate. Normale und langsame ROS haben für gewöhnlich keinen oder einen deutlich geringeren negativen Einfluss auf die Fehlerrate. Allerdings führt sehr langsame, über-artikulierte Sprache wiederum zu einer Verschlechterung der Erkennungsleistung. Methoden, die versuchen, diese Effekte unterschiedlicher ROS zu kompensieren, wurden für verschiedene Ebenen eines ASR-Systems vorgeschlagen. Diese beinhalten unter anderem Kodebuch-Adaptation im Front-End, Anpassungen der Übergangswahrscheinlichkeiten der akustischen Modelle und

[1] Es existiert keine eindeutige Definition für die Sprechrate. Neben "gesprochenen Worten pro Minute" schlagen Siegler and Stern (1995) "Phoneme pro Minute" als präziseres Maß vor.

ROS-abhängige Sprachmodell-Anpassungen (Siegler and Stern, 1995).

Variabilitäten zwischen Sprechern Zwischen-Sprecher-Variabilitäten beschreiben sprachliche Unterschiede zwischen einzelnen Sprechern. Huang et al. (2001a) führten eine Analyse von geschätzten Sprecher-Adaptationstransformationen mittels einer Hauptachsentransformation und einer "unabhängige Komponenten"-Analyse (Bishop, 2006) im Hinblick auf Variabilitäten wie Geschlecht, Akzent und Alter durch. Sie zeigten, dass die ersten beiden Hauptkomponenten der Zwischen-Sprecher-Variabilitäten dem Geschlecht und dem Akzent eines Sprechers zugeordnet werden können. Eine noch detailliertere Analyse der Unterschiede in der Physiologie des Sprachapparates von Frauen und Männern war auch Gegenstand der Arbeit von Boë et al. (2006). Dort wurde beispielsweise gezeigt, dass die Länge der Rachenhöhle von Männern die Länge der Mundhöhle innerhalb von 20 Jahren erreicht oder sogar übersteigt. Dies ist nicht der Fall bei Frauen. Im Durchschnitt ist die Vokaltraktlänge von Frauen (etwa 14.5 cm) kleiner als die bei Männern (etwa 17.5 cm), was zu einer natürlichen Variabilität innerhalb der Gruppe von Erwachsenen führt. Als zweite Hauptkomponente führen regionale und ausländische Akzente zu einer Verschlechterung der Erkennungsleistung von ASR-Systemen (Huang et al., 2001a; Lawson et al., 2003). Heimische Akzente können dabei als systematische Verschiebungen im Merkmalsraum beschrieben werden und können daher mittels akustischer Modelle, die spezifisch an heimische Akzente angepasst sind, parallel erkannt werden. Im Gegensatz dazu ist akzentuierte Sprache von nicht-heimischen Sprechern viel schwieriger zu erkennen. Einige Gründe dafür sind die Einflüsse der jeweiligen Heimatsprache und die Kenntnisse in der nicht-heimischen Sprache (Van Compernolle, 2001). Neben akzent-spezifischen akustischen Modellen werden für gewöhnlich detailierte Aussprachevarianten bei akzentuierter Spracherkennung verwendet (Lawson et al., 2003).

Das Alter ist eine weitere Variabilität, die einen Einfluss auf die Sprachcharakteristika hat und die Erkennung von Kindersprache und von Sprache älterer Menschen ist ein weiteres Feld in der Spracherkennungs-Forschung. Schwierigkeiten treten bei Kindern aufgrund der geringeren physikalischen Größe auf. Dies führt zu höheren Positionen der Resonanzfrequenzen des Vokaltrakts im Vergleich zu Erwachsenen. Auch kann eine größere Varianz bei den Formantpositionen innerhalb der Gruppe der nicht-erwachsenen Sprecher beobachtet werden. Zusätzlich weicht die Aussprache sowie das Vokabular und die Grammatik bei Kindern signifikant von denen der Erwachsenen ab. Wilpon and Jacobsen (1996) nehmen an, dass die Verschlechterung der Erkennungsleistung bei älteren Sprechern möglicherweise durch Veränderungen im glottalen Bereich und den internen Rückkoppelungsmechanismen des artikulatorischen Systems beeinflusst wird. Wilpon et. al. zeigten

Ergebnisse, nach denen ASR-Systeme für Sprecher, die jünger als 15 Jahre und älter als 70 Jahre sind, besondere Vorkehrungen im System benötigen, damit eine akzeptable Erkennungsleistung erreicht werden kann.

Variabilitäten innerhalb eines Sprechers Auf der Ebene akustischen Wellen ist ein menschlicher Sprecher kaum in der Lage, eine Äußerung exakt gleich zu reproduzieren. Die komplexe Interaktion zwischen allen Artikulatoren und den bewussten und unbewussten neuronalen Signalen führen zu Unterschieden im erzeugten Sprachsignal bei ein und demselben Sprecher. Aufgrund dessen ist die Anzahl an notwendigen Vorlagen bei vorlagen-basierten Spracherkennungs-Systemen für sprecherunabhängige Spracherkennung sehr groß. Dynamische Zeit-Verzerrungs-Methoden (engl. dynamic time warping, DTW) und Vektor-Quantisierungs-Ansätze, die im Zeitraum zwischen 1960 und 1980 entwickelt wurden, benötigen daher eine immer größer werde Menge von Vorlagen, um die Varianten ein und desselben Wortes erkennen zu können. Die Einführung der Methode der "hidden Markov"-Modelle in den Bereich der automatischen Spracherkennung während der 1980er stellte eine Lösung zur statistischen Beschreibung der Merkmals-Verteilungen dar. Diese erlaubt eine kompakte Parametrisierung, die gut mit der Menge der Trainingsdaten skaliert (Benesty et al., 2008).

Auf phonetischer Ebene ist die Koartikulation eine weitere Variabilität der Sprache. Sprache ist das Produkt von sich kontinuierlich bewegenden Artikulatoren und die Produktion eines bestimmten Lautes ist somit das Produkt einer komplexen Sequenz von artikulatorischen Bewegungen. Sprachproduktion kann als die Bewegung der Artikulatoren hin zu bestimmten Zielkonfigurationen beschrieben werden (Ladefoged, 2001). Koartikulation ist der Effekt, bei dem zeitlich benachbarte Ziele die Art und Weise der Artikulation des momentanen Zieles beeinflussen. Die Größe, in der Koartikulation auftritt, hängt auf der einen Seite von den Unterschieden zwischen den Zielkonfigurationen ab. Auf der anderen Seite hängt sie von dem zeitlichen Intervall zwischen den Zielkonfigurationen ab. So werden viele Phoneme bei schneller oder spontaner Sprache nicht vollständig realisiert.

Der Lombard-Effekt (Lombard, 1911) tritt üblicherweise unter verrauschten Umgebungsbedingungen auf. Im Allgemeinen ist dies ein Zustand, bei dem die Sprachproduktion derart geändert wird, dass die Kommunikation effektiver unter den erschwerten Bedingungen stattfinden kann (Hansen, 1996). Die Arbeit von Hansen (1996) gibt auch einen Überblick darüber, wie Stress bestimmte Sprachcharakteristika beeinflusst. Stress führt zu Veränderungen des glottalen Quellsignals, der Grundfrequenz, der Intensität, Dauer und der Vokaltrakt-Eigenschaften. Automatische Spracherkennungs-Experimente zeigen hier eine signifikante Verschlechterung der Erkennungsrate.

Variabilität existiert ebenfalls auf der kognitiven Ebene der individuellen Sprecher. Hier hängt die Bedeuting eines Wortes mit derselben Aussprache maßgeblich von dem Kommunikations-Kontext ab. Das folgende englischsprachige Beispiel stammt von Huang et al. (2001b) und demonstriert diesen Sachverhalt:

> *Mr. Wright should write to Ms. Wright right away about his Ford or four door Honda.*

In diesem Beispiel haben die Wörter "Wright", "write" und "right" alle dieselbe phonetische Realisierung. Das gleiche gilt für "Ford or" und "four door". Die Bedeutung hängt hier nicht nur von der grammatischen Rolle des Wortes ab, sondern auch von dem pragmatischen Wissen. Dies ist Wissen über den Kontext der Äußerung, über die Kommunikationspartner, deren Intentionen, deren allgemeine Werte, etc.

E.2.3. Überblick

Methoden zur Extraktion robuster parametrischer Darstellungen von einem Sprachsignal als Eingabe für ein ASR-System sind im Fokus dieser Arbeit. Verschiedene Methoden zur Verbesserung der Robustheit gegenüber einigen der oben beschriebenen Variabilitäten werden untersucht. Der Fokus der vorgestellten Methoden liegt dabei auf der Zwischensprecher-Variabilität, die durch verschiedenen Vokaltraktlängen entsteht. Ein weiterer Schwerpunkt liegt auf Methoden, die versuchen, die Robustheit gegenüber Variabilitäten wie additives Rauschen zu erhöhen. Die Hauptbeiträge dieser Arbeit können wie folgt zusammengefasst werden:

- Ein Hauptaspekt ist auf verschiedene Merkmalsextraktions-Methoden gelegt, die Verwendung von Invarianz-Tranformationen machen. Es werden Merkmalsextraktions-Methoden präsentiert, die von translations-invarianten Transformationen, die ursprünglich im Feld der Bildanalyse und -klassifikation vorgestellt wurden, Gebrauch machen. Weiterhin wird ein allgemeiner Ansatz zur Konstruktion von Invarianten, der als "invariante Integration" bekannt ist, an das Feld der automatischen Spracherkennung angepasst und eine Merkmalsextraktions-Methode basierend auf diesem Ansatz vorgestellt.
- Die Verwendung eines auditorischen Modells, welches ursprünglich entwickelt wurde, um bestimmte psychoakustische Beobachteungen zu erklären, wird mit einer der invarianten Merkmalsextraktions-Methoden kombiniert. Es wird gezeigt, dass die Merkmale, die auf diesem Modell basieren, zusätzliche Informationen zu den Merkmalen enthalten, die auf einer üblicherweise verwendeten Filterbank basieren und die Erkennungsleistung unter verrauschten Bedingungen dadurch verbessert wird.

- Zwei Methoden für die Schätzung der spektralen Effekte aufgrund von Veränderungen der Vokaltraktlänge werden in Kapitel 6 vorgestellt. Die erste Methode ist datengetrieben und erlaubt die Berechnung von Deformationen, die zwei Zeit-Frequenz-Darstellungen miteinander in ein Verhältnis setzen. Es wird gezeigt, dass die berechneten Deformationen dazu verwendet werden können, eine für gewöhnlich verwendete Methode zur Vokaltraktlängen-Normierung weiter zu verbessern. Die Zweite Methode ist modellgetrieben und schätzt die spektralen Effekte mit Hilfe eines Wellen-Reflektions-Modells des Vokaltrakts.

Die Thesis ist wie folgt aufgebaut: Kapitel 1 hat einen kurzen Überblick über die menschliche Sprachproduktion und -wahrnehmung gegeben. Auch wurden verschiedene Variabilitäten, mit denen ein Spracherkennungssystem gegebenenfalls umgehen muss, beschrieben. Kapitel 2 führt grundlegende Prinzipien und Methoden, welche durchgehend in den darauf folgenden Kapiteln verwendet werden, ein. Die grundlegende Architektur des Erkennungssystems, welches in dieser Arbeit verwendet wird, ist in Kapitel 3 beschrieben. Die verschiedenen invarianten Merkmalsextraktions-Methoden sind detailliert in Kapitel 4 beschrieben. Methoden zur Erhöhung der Rauschrobustheit eines ASR-Systems auf der Merkmalsextraktionsebene werden in Kapitel 5 beschrieben. Die Methoden zur Schätzung der spektralen Effekte aufgrund von Vokaltraktlängen-Änderungen sind in Kapitel 6 beschrieben. Eine Zusammenfassung der Methoden dieser Thesis und ein Ausblick sind im abschließenden Kapiel 7 gegeben. Informationen über die verwendeten Datensätze, die mathematische Notation und einige weitere Ergebnisse von Experimenten finden sich im Anhang.

E.3. Zusammenfassung und Ausblick

Die Erkennungsleistung eines automatischen Spracherkennungssystems erreicht derzeit nicht die menschliche Leistung. Einer der Hauptgründe dafür ist durch den hohen Grad an Variabilität der Sprachsignale gegeben, der unter praktischen Bedingungen auftritt. Die verschiedenen Arten von Variabilitäten wurden in Kapitel 1 beschrieben und gruppiert. Eine dieser Variabilitäten ist die Vokaltraktlänge und stellt einen Hauptaspekt dieser Arbeit dar. Bis heute wird Robustheit gegenüber den Effekten unterschiedlicher Vokaltraktlängen für gewöhnlich dadurch erreicht, dass entweder eine Normalisierung während oder nach der Merkmalsextraktions durchgeführt wird, oder indem die akustischen Modelle des Dekoders angepasst werden. Die Prinzipien der Normalisierung und der Adaptation wurden zuammen mit anderen grundlegenden Methoden heutiger Spracherkennungssysteme in Kapitel 2 beschrieben. Die experimentellen Konfigurationen für die Erkennungsexperimente dieser Arbeit wurden in Kapitel 3 zusammen mit Ausgangserkennungsraten von Standardmerkmalen beschrieben. Weiterhin wurden verschiedene auditorische Filterbänke im Hinblick auf die Erkennungsleistung mittels Cepstral-Koeffizienten evaluiert. Die betrachteten Filterbänke waren die Mel- und Gammaton-Filterbank, sowie the statische und dynamisch-kompressive Gammachirp-Filterbank. In diesen Experimenten führte die Gammaton-Filterbank zu den höchsten Erkennungsraten, und motiviert damit deren Verwendung bei den Methoden, die in den folgenden Kapiteln beschrieben werden. Zu Beginn des Kapitels 4 wurde eine Einführung in die Prinzipien der Invarianz-Transformationen gegeben, gefolgt von einem Überblick über verwandte Arbeiten aus dem Feld der invarianten Merkmalsextraktion für die automatische Spracherkennung. Eine zentrale Annahme der vorgestellten invarianten Merkmalsextraktions-Methoden ist, dass unterschiedliche Vokaltraktlängen zu Verschiebungen entlang der Kanalindex-Achse einer Zeit-Frequenz-Darstellung führen. Weil nicht klar ist, welche Skala zur Bestimmung der Filter-Frequenzmitten dieser Annahme am besten entspricht, wurden verschiedene Skalen, einschließlich der Mel- und der ERB-Skala, in Kapitel 4 mittels eines Spracherkennungssystems mit translationaler VTLN bezüglich ihrer Erkennungsleistung evaluiert. Die Ergebnisse zeigten, dass die Mel- und die ERB-Skala bei gleichem Filter zu ähnlichen Erkennungsraten führen und dass beide Skalen zu deutlich besseren Erkennungsleistungen führen als die Log-Skala. Zusätzlich führten die Gammaton-Filter zu höheren Erkennungsraten als Dreiecksfilter, was die Ergebnisse der Experimente aus Kapitel 3 unterstützt.

E.3.1. Invarianz-Transformationen zur Merkmalsextraktion in ASR

Drei unterschiedliche translations-invariante Methoden zur Extraktion von Merkmalen für die sprecherunabhängige Spracherkennung wurden in den Abschnitten 4.4, 4.5 und 4.6 beschrieben.

Die erste Methode basiert auf Transformationen der Klasse $\mathbb{C}T$ (Müller and Mertins, 2010b). Translation-Invarianz wird bei dieser Methode mit der Wahl von zwei kommutativen Operatoren und einer modifizierten schnellen Walsh-Hadamard-Transformation erreicht. Hier haben die Operatoren, die von der "rapid transform" bekannt sind, zu den höchsten Erkennungsraten unter den betrachteten Operator-Paaren geführt. Besonders bei Trainings- und Testmengen, die unterschiedliche mittlere Vokaltraktlängen aufwiesen, haben die CT-basierten Merkmale zu deutlich höheren Erkennungsraten als die MFCCs geführt. Diese Erkennungsraten wurden weiter erhöht, indem die Transformationen auf mehreren Skalen der Zeit-Frequenz-Darstellung durchgeführt und die Merkmalsvektoren durch korrelations-basierte "VTLI"-Merkmale von Rademacher et al. (2006) ergänzt wurden.

Die zweite Methode (Müller et al., 2009) basiert auf der sogenannten generalisierten zyklischen Transformation (engl. generalized cyclic transform, GCT). Die vorgestellte Merkmalsextraktions-Methode besteht aus zwei wesentlichen Schritten, um Translations-Invarianz zu erreichen: Als erstes werden die Spektralwerte des momentanen Frames linear mittels einer generalisierten charakteristischen Matrix (engl. generalized characteristics matrix, GCM) transformiert. Zweitens wird ein translations-invariantes Spektrum aus dem transformierten Eingangsspektrum berechnet. In dem experimentellen Abschnitt wurden verschiedene Parametersätze für die GCMs evaluiert, sowie zwei unterschiedliche Wege zur Berechnung eines translations-invarianten Spektrums betrachtet. Der Multi-Skalen-Ansatz, wie verwendet bei den CT-basierten Merkmalen, hat zur Annahme, dass Translationen entlang der gesamten Frequenzachse der Zeit-Frequenzdarstellung auftreten. Das Subframe-Schema (engl. subframing scheme), welches bei den GCT-Merkmalen verwendet wurde, hat hingegen die Annahme, das diese Translationen nur innerhalb bestimmter Frequenzintervalle auftreten. Die GCT-basierte Extraktion wurde auf diesen Abschnitten durchgeführt. Die Verwendung einzig der GCT-basierten Merkmale führte nur im Falle von unterschiedlichen mittleren Vokaltraktlängen bei Trainings- und Testdaten zu Steigerungen der Erkennungsleistung. Die Kombination mit den korrelationsbasierten VTLI-Merkmalen hingegen zeigte deutliche Verbesserungen der Erkennungsraten, auch bei gleichen mittleren Vokaltraklängen der Trainings- und Testmengen. Die Erkennungsleistung der MFCC-Merkmale wurde übertroffen. Die Erhöhung des diskriminativen Informationsgehalts durch die GCT-basierten Merkmale wird auch duch die Beobachtung unterstützt, dass die Kombination aus GCT- und VTLI-Merkmalen zu höheren Erkennungsraten führt, als es die Verwendung einzig von VTLI-Merkmalen tut.

Die erste und die zweite hier vorgestellte invariante Extraktionsmethode verwenden Invarianz-Transformationen, die speziell für Translationsinvarianz definiert wurden. Im Gegensatz dazu macht die dritte Merkmalsextraktions-Methode, welche in Kapitel 4 beschrieben wird, von einem allgemeineren Ansatz zur Berechnung von Invarianten Gebrauch. Dieser Ansatz ist bekannt als "invariante Integration" (engl. invariant integration) und stellt eine zentrale Methode dieser Arbeit dar (Müller and Mertins, 2009a, 2010a, 2011a). Die Hauptidee dieser Methode ist, (möglicherweise nichtlineare) Funktionen von Spektralwerten aus einem bestimmten zeitlichen und spektralen Kontextüber eine endliche Gruppe zu integrieren. Dadurch, dass die invarianten Integrationsmerkmale (engl. invariant integration features, IIF) auf Monomen basieren, besitzt deren Berechnung eine geringe Komplexität. Allerdings ist der Parameterraum dieser Merkmale sehr groß, so dass eine angemessene Merkmalselektion benötigt wird. Hier wurde eine Methode verwendet, die auf einem linearen Klassifikator basiert und einen Merkmalssatz fester Größe iterativ verbessert. Die Experimente haben gezeigt, dass kontext-abhängig selektierte MFCCs eine deutlich höhere Erkennungsrate als Standard- und LDA-kombinierte MFCCs in gleichen, als auch in ungleichen Trainings- und Testbedingungen aufweisen. Invariante Integrationsmerkmale können als weitere Verfeinerung angesehen werden, welche versucht, wichtige spektrale Marken innerhalb eines zeitlichen Kontexts zu finden und die Robustheit gegenüber Vokaltraktänderungen zu erhöhen. Wenn keine Sprecher-Adaptation verwendet wurde, zeigten die Experimente, dass die IIFs eine überragende Erkennungsleistung gegenüber Cepstral-Koeffizienten (sowohl MFCCs, als auch GTCCs) bei gleichen, als auch insbesondere bei ungleichen Trainings- und Testbedingungen besitzen. Bei gleichen Trainings- und Testbedingungen haben IIFs der Ordnung eins zu höheren Erkennungsraten geführt, als IIFs mit Monomen höherer Ordnung. Allerdings führte die Verwendung von Monomen höherer Ordnung bei ungleichen Trainings- und Testbedingungen zu besseren Erkennungsleistungen. Die Experimente haben gezeigt, dass die Kombination von IIFs mit MLLR und/oder VTLN deren Erkennungsrate weiter erhöht. Innerhalb der Experimente führte das IIF-basierte Erkennungssystem zu den höchsten Erkennungsraten und übertraf die Ceptral-Koeffizienten-basierten Systeme bei gleichen Trainings- und Testbedingungen auf dem TIMIT-Korpus um mehr als einen Prozentpunkt. Der letzte Teil der Experimente hat gezeigt, dass die TIMIT-basierten IIFs ebenso gut zu Worterkennungsaufgaben auf dem TIDIGITS Korpus verwendet werden können. Auch hier haben die IIF-basierten Systeme mit und ohne VTLN zu Erkennungsleistungen geführt, die höher als die der MFCC-basierten Systeme waren.

Ein Vergleich der Erkennungsleistungen der unterschiedlichen Merkmalstypen wurde am Ende von Kapitel 4 auf zwei unterschiedlichen Korpora durchgeführt. Die Ergebnisse zeigten, dass die IIFs die CT- und GCT-basierten Merkmale mit und ohne MLLR in der Erkennungsleistung auf TIMIT übertreffen. Die IIFs waren ebenfalls bei den Experimenten auf dem OLLO Korpus überragend. System-Kombinations-Experimente mittels der ROVER-Methode wurden ebenfalls am Ende von Kapitel 4 durchgeführt (Müller and Mertins, 2011c). Die drei präsentierten invarianten Merkmalsextraktions-Methoden dieser Arbeit, sowie die VTLI-Merkmale und die Standard-MFCC- und -PLP-Merkmale wurden dabei berücksichtigt. Das Back-End des Erkennungs-Systems wurde bei allen Merkmalstypen gleich gehalten. Es wurde gezeigt, dass die Kombination der erstbesten Hypothesen jedes einzelnen Systems mittels ROVER zu einer relativen Verbesserung von 11 Prozent führt.

E.3.2. Rauschrobustheit der invarianten Merkmalstypen

Rauschrobustheit der Merkmalsextraktions-Methoden bei Spracherkennungssystemen ist kritisch für viele praktische Anwendungen. Zu Beginn des Kapitels 5 wurden die beschriebenen Merkmalsextraktions-Methoden zunächst bezüglich ihrer Leistung bei verrauschten Sprachsignalen ohne weitere Verbesserungsmethoden evaluiert. Es wurde gezeigt, dass die IIFs die höchste Erkennungsleistung bis zu einem Signal-zu-Rausch-Verhältnis (engl. signal-to-noise ratio, SNR ratio) von 0 dB aufweisen und unter allen Rauschverhältnissen zu höheren Erkennungsraten führen als MFCCs. Auch wurde gezeigt, dass die Hinzunahme einer Untermenge der VTLI-Merkmale die Erkennungsleistung der invarianten Merkmale unter allen Rauschverhältnissen weiter erhöht. Anschließend wurden Methoden, die im Merkmalsraum arbeiten und typischerweise zur Verbesserung der Rauschrobustheit von ASR-Systemen eingesetzt werden, evaluiert. Die Effekte von Mittelwert-Normalisierung (MN) und Varianz-Normalisierung (VN), sowie die RASTA-Filterung, Energie-Normalisierung (engl. power-normalization, PN) und die Energie-Neigungs-Subtraktion (engl. power-bias subtraction, PBS) wurden anhand der Integrationsmerkmale ausgewertet. Die Ergebnisse zeigten, dass die Kombination von PN, PBS und MN zu den höchsten Erkennungsraten unter den betrachteten Methoden führt. Die erreichten Erkennungsraten übertrafen auch diejenigen der PNCC-Merkmale.

Der zweite Teil von Kapitel 5 untersuchte die Verwendbarkeit eines komplexeren auditorischen Modells in Kombination mit IIFs. Das "auditorische Bild Modell" (engl. auditory image model, AIM) ist ein computergestütztes Modell der menschlichen auditorischen Verarbeitungskette, welches ein Sprachsignal zu jedem Zeitpunkt zweidimensional und skalen-kovariant darstellt. Dieser Darstellungsraum wird "stabilisiertes auditorisches Bild" (engl. stabilized auditory image, SAI) genannt. Motiviert durch andere Arbeiten, die das SAI als Grundlage zur Merkmalsexteraktion für die Spracherkennung verwendeten, wurde gezeigt, wie das Konzept der invarianten Integration innerhalb des SAI angewendet werden kann. Erkennungsexperimente wurden für unterschiedliche Rauschverhältnisse auf dem AURORA 2 Korpus durchgeführt. Die Ergebnisse zeigten, dass die invarianten Merkmale, die auf dem AIM basieren, eine höhere Robustheit gegenüber Rauschen aufweisen als MFCCs, jedoch zu schlechteren Erkennungsraten führen als PNCC-Merkmale. Bei diesen Experimenten führte die Kombination aus AIM-basierten Integrationsmerkmalen mit Energie-normalisierten Integrations-Merkmalen zu den höchsten Erkennungsraten unter verrauschten Verhältnissen.

E.3.3. Schätzung der Spektralen Effekte durch Vokaltraktlängen-Änderungen

Eine zentrale Annahme der gewöhnlicherweise verwendeten VTLN mit stückweiselinearer Verzerrungs-Funktion ist die lineare Abhängigkeit zwischen den Resonanz-

Frequenzen und der Vokaltraktlänge. Diese Annahme ergibt sich aus dem Modell des Vokaltrakts als verlustfreie, uniforme Röhre. Kapitel 6 präsentierte zwei verschiedene Ansätze, die die Standard-VTLN-Methode verbessern, ohne die Annahme eines verlustfreien, uniformen Röhrenmodells zu machen.

Der erste Ansatz war datengetrieben und machte Gebrauch von elastischer Registrierung, um ein gegebenes spektrales Profil als Vorlage und ein erwünschtes spektrales Profils als Referenz mittels eines Deformationsfeldes miteinander in Bezug zu setzen (Müller and Mertins, 2012a). Die Hauptidee dieser Methode ist, die Zeit-Frequenz-Darstellung einer Äußerung derart zu transformieren, dass die transformierte Darstellung einer hypothtischen, optimalen Zeit-Frequenz-Darstellung ähnlicher ist. Elastische Registrierung wird hier verwendet, um eine kontext-abhängige, nicht-parametrische Transformation für jeden Frame zu schätzen. Die Hypothese wird dabei mittels einer standardmäßig verwendeten VTLN-Methode generiert. Daher kann die vorgeschlagene "elastische VTLN" als zusätzlicher Verbesserungsschritt innerhalb eines ASR-Systems angesehen werden. Die Ergebnisse der Erkennungs-Experimente zeigten, dass die vorgestellte Methode in der Lage ist, die Erkennungsleistung eines Monophon-Systems derart zu steigern, dass die Leistung eines Triphon-Systems ohne VTLN erreicht wird. In einem weiteren Experiment wurde eine Menge globaler Verzerrungs-Funktionen auf Grundlage der geschätzten Deformationsfelder berechnet. Anstelle der stückweise-linearen Verzerrungsfunktionen wurde diese Menge von Funktionen mit dem normalen VTLN-Ansatz, der optimale Verzerrungsfaktoren über ein Suchgitter findet, verwendet. Die Ergebnisse zeigten nur leichte Verbesserungen der Erkennungsleistung im Vergleich zu dem Standard-Ansatz. Dies zeigt die Wichtigkeit von kontextabhängiger VTLN wie es von der vorgestellten elastischen VTLN Methode durchgeführt wird.

Die zweite Methode im Kapitel 6 (Müller and Mertins, 2012b) basiert auf einem verlustbehafteten Modell des Vokaltrakts, welches ursprünglich für die artikulatorische Sprachsynthese entwickelt wurde. Das Modell kann Vokaltrakte unterschiedlicher Länge simulieren und wurde verwendet, um eine Menge von synthetischen Lauten zu erzeugen. Wie beim datengetriebenen Ansatz auch, wurde elastische Registrierung verwendet, um das Verhältnis zwischen den durchschnittlichen Spektralprofilen des gleichen Lautes unterschiedlicher Vokaltraktlängen zu beschreiben. Die geschätzten Transformationen wurden verwendet, um globale Verzerrungs-Funktionen zu schätzen, welche wiederum anstelle der stückweise-linearen Verzerrungsfunktion im Standard-VTLN-Ansatz verwendet wurden. Die Ergebnisse dieses Experiments haben leichte Verbesserungen der Erkennungsrate gezeigt, welche in ihrer Größenordung vergleichbar mit denen des datengetriebenen Ansatzes war, bei dem globale Verzerrungsfunktionen geschätzt wurden.

E.3.4. Ausblick

Während der vergangenen Dekaden wurden immer wieder ein erhebliche Fortschritte auf den verschiedenen Ebenen des Back-Ends eines ASR-Systems gemacht. Im Gegensatz dazu basieren die Front-Ends der überwiegenden Mehrheit heutiger führender Spracherkennungssysteme auf Methoden, die vor Jahrzehnten entwickelt wurden. Deren zugrunde liegende Modelle haben deutliche Defizite im Hinblick auf deren anatomische und psychoakustische Analogien. Diese Arbeit leistet einen Beitrag zu fortschrittlicheren Merkmalsextraktions-Methoden zur Spracherkennung und zeigt letztlich auch mögliche Richtungen zukünftiger Arbeiten auf.

Die Verwendung komplexwertiger Zeit-Frequenz-Darstellungen zur Merkmalsextraktion ist bei allen drei vorgestellten Merkmalsextraktions-Methoden möglich: Transformationen der Klasse $\mathbb{C}\mathcal{T}$ können auf komplexe Eingangssignal angewendet werden. Das gleiche gilt für GCT- und IIF-Merkmale. Durch die Verwendung von angemessen ausgewählten charakteristischen Koeffizienten könnte die GCT-Transformation eine komplexe Transformationsmatrix vor der Berechnung der invarianten Spektren anwenden. Die Methode der invarianten Integration erlaubt es, Invarianz gegenüber beliebigen Transformationsgruppen zu erreichen. In dieser Arbeit wurde die Gruppe der Translationen betrachtet, womit bereits deutliche Steigerungen der Erkennungsleistung beobachtet werden konnten. Anstelle von Translationen könnten andere Transformationsgruppen die spektralen Effekte aufgrund von unterschiedlichen Vokaltraktlängen besser beschreiben. Diese Transformationen könnten mit dem Integrationsansatz verwendet werden. Eine kritische Eigenschaft aller drei vorgestellten Extraktionsmethoden ist die hohe Dimensionalität der resultierenden Merkmalsvektoren. Für die Experimente in dieser Arbeit wurde die LDA zur Dimensionsreduktion verwendet. Dieser Ansatz stellt einen Kompromiss zwischen rechnerischer Effizienz und Erkennungsleistung dar. Im Hinblick auf die Forschungsgemeinde im Bereich der Spracherkennung existieren Arbeiten, die verschiedene weitere Dimensionsreduktions-Methoden verwenden (zum Beispiel, Kumar and Andreou, 1998; Saon et al., 2000; Heckmann and Gläser, 2011), welche auch hier zu weiteren Verbesserungen der Erkennungsleistung führen könnten. Auch haben informationstheoretische Transformationen (Torkkola, 2003; Ozertem et al., 2006; II et al., 2006) vielversprechende Fortschritte im Vergleich zu traditionellen Transformationen wie die PCA oder LDA gezeigt. Deren Anwendung für die Selektion von Integrationsmerkmalen oder zur Dimensionsreduktion stellen interessante Alternativen im Kontext dieser Arbeit dar. Eine andere Ebene der Merkmalsextraktion für die Spracherkennung stellt die Berechnung einer Zeit-Frequenz-Darstellung dar. Die Mel- und Gammaton-Filterbänke, welche in den Standard-Extraktionsmethoden verwendet werden, können viele psychoakustische Beobachtungen nicht erklären. Dennoch wurde in dieser und anderen Arbeiten (zum Beispiel, Walters, 2011) gezeigt, dass die Verwendung von komplexeren Modellen, die eben diese Beobachtungen erklären, nicht unbedingt zu besseren Erkennungsleistungen führt. Es wäre interessant, die Gründe für diese Diskrepanz in zukünftigen Arbeiten weiter zu untersuchen. Die Ergebnisse dieser Untersuchungen könnten dazu führen, dass die Potentiale der komplexeren auditorischen Modelle besser ausgenutzt werden können. Beispielweise verwenden die AIM-IIFs, wie in dieser Arbeit beschrieben, eine Gammaton-Filterbank zur Simulation der Bewegungen der Basilarmembran. Eine Optimierung der Parameter der komplexeren dynamisch-kompressiven Gammachirp-Filterbank könnte die Erkennungsleistung der AIM-basierten IIFs weiter steigern.

Die Rauschrobustheit der präsentierten Merkmalsextraktions-Methoden wurde innerhalb dieser Arbeit für additiv gestörte Äußerungen evaluiert. Zukünftige Arbeiten im Feld der Rauschrobustheit für invariante Extraktionsmethoden könnten Anwendungsfälle betrachten, bei denen der Lombard-Effekt beobachtbar ist und dadurch ein höherer Grad an Realismus erreicht wird. Speziell die IIFs haben aufgrund ihrer allgemeinen Parametrisierung ein hohes Maß an Flexibilität. Eine Möglichkeit davon Gebrauch zu machen, wäre eine adaptive Parametrisierung derart, dass eine Robustheit gegenüber weiteren Variabilitäten als die Vokaltraktlänge erreicht wird.

Die vorgeschlagene "elastische VTLN" als Verbesserung der Standard-VTLN ist ein datengetriebener Ansatz zur Schätzung der spektralen Effekte unterschiedlicher Vokaltraktlängen. Dieser Ansatz hat in den hier beschriebenen Experimenten zu großen Verbesserungen der Erkennungsleistung geführt. Die Wahl vom Distanzmaß und vom Regularisierer für diesen Ansatz haben bedeutende Auswirkungen auf die Lösungen dieser Methode. Es wurde in den beschriebenen Experimenten gezeigt, dass die hier getroffene Wahl vielversprechend ist. Allerdings könnten weitere Experimente mit anderen Distanzmaßen und Regularisierern zeigen, ob es eventuell noch vorteilhaftere Maße gibt. Eine weitere Verfeinerung der Registrierungsmethode wäre die Einführung eines weiteren Strafterms. Die hier verwendete Zielfunktion berücksichtigt nicht den Energieerhalt (im Hinblick auf die Spektralwerte) während der Berechnung der Transformation. Ein entsprechender Strafterm könnte hier diese Funktion übernehmen. Wegen der Normalisierung während der nachfolgenden Merkmalsextraktion kann in dieser Arbeit davon ausgegangen werden, dass das Mißachten der Energieerhaltung vernachlässigbar ist. Eine genaue Analyse und Verbesserung dieses Sachverhalts könnte allerdings zu Verbesserungen der Erkennungsleistung führen. Die Verwendung größerer Korpora würde es erlauben, Referenz-Triphon-Modelle zu trainieren. Auch könnte die vorgestellte elastische VTLN-Methode für mehrere Durchgänge angewendet werden und so zu einer weiteren Steigerung der Erkennungsrate führen.

Die modellgetriebene Methode zur VTLN kann in verschiedenen Aspekten weiterentwickelt werden: Artikulatorische Modelle hängen maßgeblich von Ergebnissen von Magnetresonanztomographie (MRT) Studien ab, mittels derer die Vokaltraktkonfigurationen einzelner Laute bestimmt werden. Untersuchungen durch weiteren MRT-Studien würden zu weiteren Verfeinerungen und einer höheren Generalisierbarkeit der Modelle führen. In der momentanen Implementierung werden unterschiedliche Vokaltraklängen durch die Wahl einer unterschiedlichen Anzahl an Röhrenelementen und linearer Interpolation der Durchmesser- und Basis-Funktionen $\Omega(k)$, $\Phi_1(k)$, $\Phi_2(k)$ realisiert. Es wurde beispielsweise von Boë et al. (2006) gezeigt, dass das Verhältnis der Rachenhöhlen-Länge zu der Mundhöhlen-Länge bei Männern und Frauen unterschiedlich ist. Daher würde eine Modellierung, die diesen Umstand berücksichtigt, zu einer realistischeren Simulation führen. Auch existieren andere artikulatorische Sprachsynthese-Modelle, welche für einen modellgetriebenen Ansatz, wie hier beschrieben, verwendet werden könnten. Ein bestimmtes Modell, von dem gezeigt wurde, dass es in der Lage ist natürlich klingende Sprache zu erzeugen, wird in der Arbeit von Birkholz (2005) beschrieben. Allerdings basiert auch dieses Modell auf den Ergebnissen von MRT-Studien.

Wie bereits bei dem datengetriebenen Ansatz weiter oben beschrieben, könnte eine andere Wahl des Distanzmaßes oder des Regularisierers zu besseren Transformationsschätzungen führen. Wie beobachtet in den Experimenten von Abschnitt 6.1.3, hat eine nicht-globale Anwendung der Verzerrungsfunktionen einen kritischen Einfluss auf die Erhöhung der Erkennugsleistung eines Erkennungssystems. Daher verspricht eine phonemabhängige Schätzung und Anwendung der Verzerrungs-Funktionen die Normalisierungs-Eigenschaften dieses Ansatzes weiter verbessern zu können.

Index

Besides a listing of elementary notions within this work abbreviations together with their meaning are listed here.

Curriculum Vitae

Personal Details

Name	Florian Müller
Date of birth	07.07.1982
Born in	Lübeck, Germany

Experience

01.2008 – 02.2013	Research associate at the Institute for Signal Processing, University of Lübeck
03.2007 – 12.2007	Student trainee and graduation thesis at Philips Research Laboratories, Digital Imaging Group, Hamburg

Education

08.2002 – 12.2007	Study of Computer Science at the University of Lübeck
12.2007	Diploma degree in computer science Title of the thesis: *"Lung nodule growth measurement with model-based 4D-segmentation and elastic registration"*

Secondary School

1995 – 07.2001	Ostsee-Gymnasium Timmendorfer Strand